From **Shelf** to **Startup**

BREWSTER HOUSE

PUBLISHED BY BREWSTER HOUSE

Oxford, Mississippi

© 2025 Parker Brewster. All rights reserved.

All 'characters' in this book are fictitious, and any resemblance to actual persons, living or dead, is purely coincidental.

No part of this book may be reproduced, distributed, or transmitted in any form or by any means without the prior written permission of the publisher, except in the case of brief quotations used in reviews, articles, or scholarly works.

All advice, guidance, and commentary contained in this book reflect the author's personal opinions, experiences, and observations. Nothing in this book should be interpreted as legal, financial, or corporate advice. Readers should seek qualified professional counsel before making decisions in these areas.

Authored by: Parker Brewster

Secondarily authored by: Eden Wallace

Book designed and illustrated by: Sandra Rakasova

Paperback ISBN: 979-8-9931370-0-1

First Edition, 2025

Table of Contents

Message IX
From the Author

Aknowledgement: XI
A Debt to Biodesign

Part 0 (Starting From Zero) 1
Chapter 0.1: You Don't Need to Be 'Ready' to Start 2
Chapter 0.2: Finding the Time When You Don't Have It 4
Chapter 0.3: Navigating Doubt, Imposter Syndrome, and Skepticism 5

Part I (Vision and Foundation Building) 8
Chapter 1: The Playbook You've Been Waiting For 9
Chapter 2: Why Student Innovation Gets Stuck 14
Chapter 3: What You're Building: Startups, Ventures, or Systems? 19
Chapter 4: Building Your Network with Intention 29

Part II (Problem Discovery) 50
Chapter 5: Where to Find Real Problems 51
Chapter 6: Writing the Right Problem 66
Chapter 7: From Many to One: Funneling Your Need Statements 81

Part III (Landscape Mapping) 93
Chapter 8: Mapping the Landscape BEFORE You Build 94
Chapter 9: Understanding the Terrain 109
Chapter 10: Markets: Who Pays, and Who Benefits? 125

Part IV (From Idea to Form) 141
Chapter 11: Translating Needs into Concepts 142
Chapter 12: Designing for Real Constraints 158
Chapter 13: Prototyping Like a Builder 173

Part V (Testing the Right Things) 183
Chapter 14: Testing For Value and Usability 184

Part VI (Customer Discovery with Confidence) 197

Chapter 15: Customer Discovery with Confidence 198
Chapter 16: Refining What You've Got. 224

Part VII (Understanding the Terrain Ahead). 241
Chapter 17: Intellectual Property for First-Time Founders 242
Chapter 18: Clinical Commercialization 101 (SKIPPABLE for
               Non-Clinical Readers). 258
Chapter 19: Translating Clinical Knowledge into Business Strategy
               (SKIPPABLE for non-clinical readers). 271
Chapter 20: Common Business Models in Healthcare and Beyond
               (For All Founders) 281

Part VIII (Pathways to Build Forward) 291
Chapter 21: From Prototype to Pathway: Venture or Licensing? 292
Chapter 22: Startup Basic for Builders 304
Chapter 23: Strategic Planning and Risk Readiness 315
Chapter 24: The Science of Risk Management 324

Part IX (Structuring for Success). 336
Chapter 25: The Lean Canvas as a Living Tool. 337
Chapter 26: Pitching with Purpose. 348
Chapter 27: Building a Team and Leading It 358

Part X (Financial Fluency for Founders). 376
Chapter 28: Demystifying Funding Pathways. 377
Chapter 29: How to Read a Term Sheet. 389

Part XI (Fundraising Strategy and Execution) 402
Chapter 30: Building a Fundraising Strategy. 403
Chapter 31: Money Without Dilution. 415
Chapter 32: What do VC's Really Want?. 426

Part XII (Building Forward) 446

Challenge Index 451

Message

From the Author

When I began pulling together the frameworks that would become this book, I wasn't writing for publication. I was building an internal playbook for Biomedical Think-Tank, a venture studio I founded to help students transform overlooked clinical problems into actual patient-facing companies. But something happened as that internal document grew. Mentors, fellows, partners, and peers kept asking for it, and they'd come back with a variation of the same message: "You need to make this public." Eventually, I listened.

This book is the result of that call to action. It's built on years of trying, failing, iterating, and occasionally getting it right. I've launched and co-launched multiple ventures. I've served in executive roles at medtech startups, won national pitch competitions, and mentored over 20 student-founded ventures from first idea to funding. I've been accepted into and trialed by multiple nationally ranked startup accelerators, gaining firsthand insight into what these programs do well and where many fall short. My work spans due diligence and strategic roles at five VC firms and investment groups, alongside two different fellowships through the NIH (including a translational science fellowship with NCATS), as well as programs at Harvard, Tulane, and Babson. I've built devices from scratch in shared academic labs, supported startups through FDA mapping and IP strategy, and led commercialization efforts rooted in clinical validation. Through it all, I kept wondering: why don't more students get this kind of structure earlier?

That's what this book is for.

You'll find lessons rooted in design thinking, translational science, and venture creation, but stripped of jargon and fluff. This is not an academic exercise. It's a builder's guide. Most of all, this book is a love letter to untapped student potential. I believe we're sitting on a mountain of innovation that quietly evaporates each year because the systems to capture it simply don't exist. If even a fraction of that energy were redirected toward problems worth solving, the healthcare system would look different in a decade.

My hope is that this book helps make that happen.

—Parker Brewster
Founder, Biomedical Think-Tank, Inc.

Acknowledgement
A Debt to Biodesign

Before anything else, this book owes a sincere and enduring debt to the **Stanford Biodesign Textbook**. Few works have so thoroughly shaped how I think about innovation, problem discovery, and the path from unmet need to patient impact. It is, in every sense, a cornerstone of modern health innovation education that I've studied, revisited, and taught from more times than I can count.

For me personally, Biodesign wasn't just a textbook. It was a mentor in print. As a student founder building my first companies, and later as someone creating a venture studio built to help other students do the same, it gave me the scaffolding I needed to move from intuition to structure. It introduced me to ideas that felt like secrets, only they weren't secrets, they were frameworks, generously shared by the very best in the field. But I want to be very clear: **this book is not a replacement for Biodesign. It's not an alternative, and it's certainly not a competitor.** This is a supplement. A companion. A reflection of lived experience filtered through the lens of a student builder who started with no funding, no institution, and no clear path. It was written for those who are closer to the beginning than the end, those still sitting in undergrad lecture halls or researching between shifts at the hospital.

Everything on these pages is a product of my own journey and the journeys of the students and early-stage founders I've worked with through the Biomedical Think-Tank. **Nothing here reflects the views of Stanford University or the authors of the Biodesign textbook.** If anything, I hope this work sends more readers to their doorstep. If you're serious about healthcare innovation, **read Biodesign**. If you're a student wondering how to start building without a lab, a budget, or a map, **this book is for you**

—Parker Brewster
Founder, Biomedical Think-Tank, Inc.

From **Shelf** to **Startup**

A PLAYBOOK
FOR TRANSFORMING
UNMET NEEDS
INTO THRIVING VENTURES

Parker Brewster

BREWSTER HOUSE

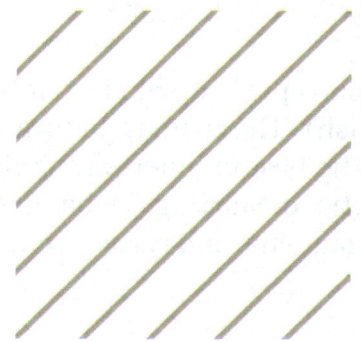

Part 0

Starting From Zero

CHAPTER 0.1: YOU DON'T NEED TO BE 'READY' TO START

INNOVATION AS A MUSCLE, NOT A MOMENT

The biggest myth in entrepreneurship is that there's a perfect time to begin. There isn't. The truth is, innovation isn't a lightning strike; it's a muscle. The best founders aren't the most prepared, they're the most practiced. Starting is what makes you ready. It builds the skill set, stamina, and resilience that all future steps require.

At our studio, Biomedical Think-Tank, we've seen students with no technical background, no money, and no institutional support build real ventures. What they shared wasn't privilege or polish. It was courage, curiosity, and an absurd willingness to keep trying.

WHY YOUR BACKGROUND, IDENTITY, OR LACK OF EXPERIENCE IS NOT DISQUALIFYING

If you're the first in your family to pursue higher education, if English isn't your first language, if you don't have mentors or industry ties... You are not behind. By navigating ambiguity, overcoming these challenges, and advocating for yourself, you have the tools already needed to succeed in this space.

Innovation ecosystems often reward polished speech, elite networks, and credentials that can feel out of reach. Entering these spaces without them is undeniably difficult: doubts creep in, and the comparison to peers who "fit the mold" more neatly can feel discouraging. Acknowledging this head-on is important, because it is a major challenge to walk into a room where pedigree seems to overshadow substance.

But those doubts are not the whole story. What you bring to the table is not a lack, but a different kind of strength. If your background has forced you to navigate ambiguity, to advocate for yourself, or to solve problems without a ready-made playbook, those same habits are exactly what early innovation requires. The market ultimately does not reward polish for its own sake: it rewards the ability to spot problems others overlook and to pursue solutions others dismiss.

That tension is real, and you will need to push against the bias toward surface-level credibility, but you also hold unique advantages. Lived experience can illuminate unmet needs invisible to those who have only followed conventional paths. Perseverance can sustain projects through uncertainty in ways that polish alone cannot. When you begin to see your own history not as a deficit but as a toolkit, the doubts remain present but no longer dictate the outcome.

EXAMPLES OF LOW-RESOURCE FOUNDERS WHO BROKE THROUGH

- **Airbnb:** Started by renting out air mattresses during a design conference because they couldn't afford rent. Later funded their first run by selling cereal boxes.

- **Spanx**: Sara Blakely had $5,000 in savings and no experience in fashion. She built a billion-dollar brand by solving a personal discomfort.

- **Shopify**: Born from a small snowboarding shop that was frustrated by the lack of good online store options.

These weren't born from capital or connections. They were born from building anyway.

CHAPTER 0.2: FINDING THE TIME WHEN YOU DON'T HAVE IT

BUILDING WHILE BALANCING SCHOOL, JOBS, OR FAMILY

Almost every student founder we work with is managing multiple full-time responsibilities. Trying to build something on top of that can feel overwhelming, and the search for "balance" often becomes another source of stress. What matters more than balance is rhythm.

Rhythm does not mean working on your idea every single day in a rigid routine. It means setting a sustainable cadence that you can actually keep, one that matches your energy, obligations, and ambition. For some, that might mean short daily bursts of effort; for others, it might mean carving out larger blocks of time a few days a week. What matters is not the exact schedule but the fact that it repeats and compounds.

Progress becomes possible when you stop measuring yourself against an ideal of constant hustle and instead build systems that make progress inevitable. Even small, irregular steps accumulate when they are guided by a deliberate rhythm. The goal is to align your commitment with a pattern of work that fuels you rather than drains you. When you create that rhythm, you can move forward even while carrying the weight of school, jobs, or family.

TIME-BLOCKING, ENERGY MANAGEMENT, AND MOMENTUM PSYCHOLOGY

- **Time-blocking**: Set recurring windows for focused work (e.g., Sunday evenings, early mornings, lunch breaks).

- **Energy over time**: Match your hardest tasks to your most alert hours. Save busywork for your off-times.

- **Momentum matters**: Even 30 minutes of progress a week builds identity and inertia. Document what you've done and what's next; this helps you jump back in.

※ **BMTT Tip:** Sprints Over Slogs

Most student founders don't have the luxury of working full-time on their startup. Instead of slow, ongoing work, try to build in high-energy bursts.

⊗ For example:

- One weekend hackathon to make a prototype.

- A one-week sprint to send out 20 cold emails.

- A late-night brainstorm session with friends to generate ideas.

Short, focused efforts often move the needle more than scattered daily attempts.

CHAPTER 0.3: NAVIGATING DOUBT, IMPOSTER SYNDROME, AND SKEPTICISM

WHAT TO DO WHEN PEERS OR FACULTY DON'T TAKE YOU SERIOUSLY

Some students hear, "That's cute," or "Come back when you're serious." If this happens to you, don't internalize it. Most systems are not designed to encourage early-stage creativity, especially when it comes from people outside of traditional power circles. At the same time, it helps to recognize that skepticism often stems from uncertainty about your intentions. Faculty and peers may not know if you are casually tossing around ideas or if you are committed to testing whether a real need exists.

One of the most effective ways to bridge that gap is to frame your conversations clearly. Instead of leading with your solution, lead with the problem you are exploring and ask for their expertise in judging whether it is substantial enough to pursue. For example: "I'm trying to understand if there is a real clinical need here, and I value your perspective. Do you think this problem is significant enough to warrant building around?" This positions you as someone who is serious about investigation, not just speculation.

What you can do:

- Document everything. Let traction speak where titles can't.

- Find your tribe: Join entrepreneurship clubs, pitch competitions, or incubators where others are in your shoes.

- Build small wins: A working prototype, a published blog post, or a cold email response is more powerful than abstract ambition.

- Frame conversations as strategic asks: Signal that you are in the discovery phase and invite faculty to weigh in on the size of the problem, not just the feasibility of a solution.

By shifting how you communicate, you not only defuse dismissive comments but also invite your critics into the process. When people see that you are serious about validating need, they are more likely to take your effort seriously and even contribute to its growth.

HOW TO BUILD CONFIDENCE WITHOUT ARROGANCE

Confidence doesn't mean pretending to know everything. It means:

- Being comfortable saying, "I don't know yet, but I'm figuring it out."

- Taking ownership of your learning curve.

- Asking for feedback early and often.

Arrogance says, "I know enough." Confidence says, "I trust myself to learn what I need."

Remember, even the most successful founders started with nothing but an idea. What matters most is what you do next.

Part I

Vision and Foundation Building

CHAPTER 1: THE PLAYBOOK YOU'VE BEEN WAITING FOR

INTRODUCTION

Let's begin with a simple truth: most great student ideas never make it off the shelf.

Capstone projects, half-built prototypes, research posters, and design challenges, each with real potential, are routinely left behind. Not because they aren't good enough. Not because students don't care. But because the path forward is unclear, the support systems are limited, and no one ever handed them the tools to build something real.

We created Biomedical Think-Tank (BMTT) after watching this happen far too often. Students would identify a problem worth solving, sometimes even design a solution, only to hit a wall. They didn't know how to validate what they had, how to build a team, or how to move their idea forward once the class or semester ended. The energy was there. The ideas were there. What was missing was a system.

You likely don't have a BMTT near you yet. So until you do, this book is your system.

WHO THIS IS FOR

This playbook was written with students in mind, but not only students. It is for anyone who can see a problem clearly and wants a path to build something real.

- The first-year premed who already sees broken processes in care.

- The senior engineering student with a working prototype and no idea what to do next.

- The graduate student whose thesis holds more potential than they have been led to believe.

- The early-career professional who has lived a problem and feels like an outsider to the startup world.

- The community college student, the international student, the nontraditional student, and the caregiver with limited time but strong motivation.

- The intrapreneur inside a clinic, lab, or company who wants to fix a workflow instead of founding a company.

If you do not know where to start, or you feel like you are missing the network, the funding, or the credentials to begin, this book is for you.

WHAT THIS BOOK IS (AND ISN'T)

This is not a highlight reel of inspirational stories or a surface-level tour of entrepreneurship. There are plenty of those. This is a practical, high-utility playbook that walks you from need discovery to validation, early builds, and first traction. You will find frameworks, scripts, checklists, and examples that you can use the same day you read them.

This book is not theory first. It is action first. You will learn how to define a real problem, test whether it matters, map the landscape, and choose a path that fits your constraints. You will also learn how to avoid the most common traps that stall student projects: building too soon, scoping too broad, and chasing polish instead of value.

This is a builder's companion. Read a section, do the exercise, move forward. Come back when you get stuck. Use the tools again when your context changes.

HOW TO USE THIS BOOK

Work through it in sprints. Each chapter includes a concrete output: a question set, a template, a field script, or a decision tool. Save your outputs from challenges in one place. You will reuse them.

- If you have one hour a week: do one exercise per week and log the next action before you stop.

- If you have a weekend: pick a chapter with a field activity, run five interviews or one small test, then document what you learned and what it changes.

- If you have a team: divide roles by chapter outputs or challenges rather than job titles so momentum compounds.

This is not meant to be read once. It is meant to be worked.

WHY I WROTE THIS

The numbers matter, but the people are what pushed me to write.

I have watched strong student prototypes get boxed up at the end of a semester and never leave the lab. Months later, a caregiver or a physician would email to ask whether that student project ever moved forward because a patient could use it. I have opened messages from families who are tired, scared, and hopeful enough to ask a stranger whether a better tool exists. I have stood in hallways with clinicians who shrug at workarounds that should not be normal and who say, without anger, that nothing will change before the next shift.

Those moments are why this book exists. The loss is not purely academic. It is human.

I have also seen what happens when a student is given a clear, practical path. Small tests lead to real traction. A single validated workflow change improves a unit's day. A careful pilot makes a physician an ally. A community college student with no obvious advantages builds a solution that a large system chooses to adopt. None of this requires permission to begin. It requires structure, language, and a way to move.

I could not keep watching projects collect dust while patients and clinicians kept asking for progress. This book is my response. It gives you the path I wish more students had on day one, so that fewer good ideas are lost to the shelf and more real problems get solved.

STARTING ALONE IS NORMAL, AND YOU SHOULDN'T WAIT

Across disciplines, we see the same hesitation. Business students wait for "the technical cofounder." Engineers and scientists wait for "the CEO type." Everyone waits for perfect timing and a full team.

Start with what you have. That is how nearly every real venture begins.

Why you should not wait:

1. Clarity comes from contact. Short cycles of interviews, sketches, and tiny tests reveal what matters. You cannot think your way into that signal without touching the problem.

2. Semester clocks erase momentum. If you wait for the perfect conditions, the calendar will do the killing for you.

3. Networks form around motion. People say yes to specific next steps. You find mentors, collaborators, and funders

after you show the first evidence that you will move with or without them.

What starting alone looks like in practice:

- Most simply: START THE CHALLENGES in this playbook

- Pick a narrow slice of the problem to learn about this week. Run five stakeholder conversations using a script. Log exact quotes, not summaries.

- Write one solution-independent need statement. Pressure-test it with two people who work in that context.

- Build the smallest thing that lets you learn. That might be a clickable mockup, a taped-together physical model, or a workflow script you role-play in a sim lab.

- Set a cadence that you can keep. Two hours on Wednesday night and a three-hour block on Sunday can beat a tired daily grind. Consistency over theatrics.

You do not need to know everything. You do need to begin. This book will help you turn that first motion into a repeatable path.

A FINAL WORD BEFORE WE BEGIN

The problem isn't that students aren't ready to build. It's that no one built the runway.

This book is here to change that. It's here to hand you the tools, the frameworks, and the confidence to start. You don't need to be an expert. You don't need to wait. You need a clear place to begin, and now you have one.

Let's get to work.

CHAPTER 2: WHY STUDENT INNOVATION GETS STUCK

The Shelf Syndrome: Underfunded, Unsupported, and Unfinished

This chapter is not about solutions. It is about understanding the problem.

Every year, thousands of student-led projects begin with promise. A clinical insight scribbled in a notebook. A device designed for a senior capstone. A health tech prototype that almost made it to testing. These ideas are thoughtful, relevant, and often technically sound. But most of them go nowhere.

They are not abandoned because students are lazy, unmotivated, or naive. They are lost for much more systematic reasons. Most student innovations begin with energy and urgency but then quietly fall off due to invisible obstacles that few people talk about and even fewer are trained to navigate.

We call this **Shelf Syndrome:** a three-part pattern that explains why otherwise viable ideas get underfunded, unsupported, and ultimately left unfinished.

We do not present this to discourage you. Quite the opposite. Understanding these patterns will help you recognize them in your own path, and the chapters that follow will show you how to avoid repeating them. But before we move into strategy, we must confront the structure that is currently failing early innovators.

This chapter is about digesting that reality: clearly and honestly.

1. UNDERFUNDED

The first and most obvious barrier to student innovation is money.

The costs of building anything real go far beyond a laptop and an idea. Prototyping, user interviews, travel to meet stakeholders, purchasing materials, filing for intellectual property, or even getting early legal or regulatory advice... these things add up quickly. Most students do not have disposable income to cover those needs, and many do not even know where to begin looking for outside support.

University grants, when available, are often limited to a small number of students, tied to a particular department, or so competitive that most applicants get turned away. Prize money from pitch competitions can help, but it usually comes without follow-up support or structure.

External funding sources like NIH or NSF grants sound promising, but they are often inaccessible without mentorship, institutional support, or guidance on how to submit. Many students don't even know those mechanisms exist.

Meanwhile, traditional startup investors generally avoid student-led teams unless they come from high-prestige universities or have some kind of pre-existing validation. Most students are never taught how to structure or present their ideas in ways that resonate with funders. As a result, the funding gap becomes self-reinforcing: those without access are told to wait, or to pursue their ideas "someday."

It is important to acknowledge that many people with great ideas will never get to test them because they were never given the minimum resources to try.

2. UNSUPPORTED

Funding is only part of the picture, and as you will see in the chapters ahead, it is often one of the more straightforward limitations you will face as a founder. The deeper and more persistent challenge is structural.

Students are often trained in technical knowledge or business theory, but even then, the nature of an undergraduate education is that you are rarely taught actionable specifics. Coursework rarely includes how to conduct real customer discovery, how to scope a regulatory strategy, how to form a legal entity, or how to build an interdisciplinary team.

Curricula are designed to produce skilled employees, not founders. The academic system rewards clarity and correctness, not ambiguity, risk-taking, and iteration.

Even at schools with innovation hubs or entrepreneurship initiatives, the resources are frequently siloed. Engineering students may have access to labs but no understanding of the regulatory implications of their design. Business students may know how to build financial models but not how to build the product they are modeling. Medical students often live closest to the problem space but do not know how to structure a solution.

This fragmentation results in well-intentioned students being told to "go talk to someone else" again and again until momentum dies. They might be invited to events or encouraged to join a mailing list, but when it comes time to take the next step, they are still on their own.

The irony is hard to ignore. Young students and early-career professionals are often the most well-positioned to notice unmet needs. They are embedded in broken systems. They bring fresh perspectives and a willingness to question established processes. Yet, as a society, we offer them the least support when they try to build.

They are asked to be creative but not disruptive. They are praised for identifying problems, but rarely given a pathway to solve them. We act surprised when their ideas don't advance, but we have not created the conditions for them to succeed.

3. UNFINISHED

The final layer of Shelf Syndrome is time and momentum. Even students who have identified a strong need, developed a potential solution, and found modest resources still face one of the most difficult hurdles: finishing.

Most academic environments are built around semesters. A class project has a fixed end date. A competition has a deadline. A grant has a defined submission window. But innovation rarely follows such clean boundaries.

It is common to see students build something promising during a course or sprint, only to hit a wall once the structure disappears. Classes end, team members graduate, internships or exams take priority, and what felt urgent becomes background noise.

Even when someone wants to keep going, they may not know how. One of the most common forms of friction we see is when a group project has startup potential but no clarity around ownership. A student might hesitate to move forward out of fear that others on the team will feel excluded or entitled to equity. Conversations about co-founders, contributions, and accountability are rarely taught and often avoided, which means good ideas are lost not because of conflict, but because of uncertainty.

Momentum is fragile. Without a roadmap, even the most passionate student can lose direction after just a few missed weeks. And once a project is shelved, reviving it becomes exponentially harder.

THE BROADER CONSEQUENCES

When a student walks away from a promising idea, the loss is not theirs alone. It is a missed opportunity for everyone.

Each shelved project might have held the key to solving a real-world health challenge, improving a clinical workflow, or creating a more equitable access point for underserved populations. But because there was no follow-up structure, that opportunity fades into background noise.

Multiply that by every university in the country and you begin to see the magnitude of what is being lost. We are not just underleveraging student creativity. We are actively wasting it.

Students are not only capable of innovation. In many cases, they are the best positioned to lead it. They are not bound by institutional inertia. They are exposed to real-world problems earlier and more frequently than ever before. They are often more diverse, more adaptable, and more resourceful than the conventional startup founder archetype.

But without systems, support, and language for what comes next, they walk away.

CHAPTER 3: WHAT YOU'RE BUILDING: STARTUPS, VENTURES, OR SYSTEMS?

Understanding your trajectory and what success can look like

Before you dive deeper into problem discovery and prototyping, it's worth zooming out for a moment to ask a deceptively simple question:

WHAT ARE YOU ACTUALLY TRYING TO BUILD?

It's easy to default to the word "startup." In fact, the term has become so overused that it often loses meaning. But the path forward is not always as straightforward as founding a company. Sometimes what you're building is a single product with a narrow focus. Sometimes it's a larger system, or a solution meant to integrate into existing infrastructure. And in many cases, the right move is not to build a standalone company at all, but to embed your idea into someone else's ecosystem through licensing, partnerships, or a strategic handoff.

Each of these outcomes can be successful. But they each require different forms of commitment, different timelines, and different metrics for progress.

This chapter will help you clarify what direction you're aiming for, what each trajectory entails, and how to think about success in a way that aligns with your strengths and goals: not someone else's expectations.

Let's Define the Terms

Before we go further, let's define what we mean when we use terms like *startup, venture, or system-level intervention*. These terms are often used interchangeably, but the distinctions matter.

Startup

A startup is an early-stage company built to test, validate, and eventually scale a specific product or service. Startups are typically high-risk and high-reward: they require building a business from scratch, managing uncertainty, and often raising outside capital to move fast. Startups may begin with a single prototype or insight, but the eventual goal is to build a repeatable model for value creation that can survive independently.

Key features of a startup:

- You are responsible for execution
- You likely need co-founders or early team members
- You will need to think about market entry, revenue, and IP
- You may seek angel investment, venture capital, or non-dilutive grants

Venture

A venture is a broader term that includes startups, but also includes initiatives that may be housed within other entities: for example, a new commercialization pathway inside a university, or a research initiative spun out of a hospital that exists as a project, not a company.

At BMTT, we often refer to "ventures" when a solution has commercial or translational potential but has not yet become (or may never become) a standalone company. This might be because the idea is better suited for a licensing deal, an acquisition, or integration into another company's portfolio.

Key features of a venture:

- You are building toward a commercial outcome, but not necessarily full independence

- You might partner with another studio, lab, or external founder group

- You still need early validation and framing, but not necessarily a pitch deck on day one

System

A system-level intervention is broader than a product or company. You may be designing a new workflow, proposing a better data model, altering how care is delivered, or changing the user interface of a process that touches multiple stakeholders. These innovations often emerge from students with firsthand experience navigating broken systems: med students noticing inefficiencies in documentation, undergrads seeing disparities in care delivery, engineering students identifying redundancy in how tools are sterilized or reused.

Systemic innovation is harder to prototype, but potentially more powerful. It often requires working with hospitals, government agencies, insurers, or academic institutions. The payoff may be slower and harder to measure, but the impact can be transformative.

Key features of a system-level project:

- You are trying to change or influence a workflow, not just a device or app

- Success may depend on integration and stakeholder alignment

- You may need to think like a policymaker or operations leader, not just a product builder

WHY THIS DISTINCTION MATTERS

Too often, student innovators assume that "building something" must mean launching a startup. But not all problems require a new company. Not every insight is best served by raising venture capital or hiring a team. In fact, many good ideas fail precisely because they are pushed into the wrong model too early.

Part of our goal at BMTT is to help you identify what *kind* of solution you're building and how to get it to the right destination, not just the flashiest one.

A well-crafted diagnostic workflow that gets adopted by a regional hospital network might have more patient impact than a $3 million seed-stage startup with a polished slide deck but no clinical buy-in. A software patch that gets licensed to an EHR company may not come with Forbes headlines, but it may quietly improve outcomes for tens of thousands of patients a year.

That's real impact. That's real innovation. And that's what we care about.

YOU CAN START SMALL AND STAY STRATEGIC

Another mistake we see is students thinking they need to solve the whole problem at once. You don't. In fact, you shouldn't.

You might start by designing a single-use tool for a narrow problem. As you validate and explore, that tool may become part of a larger system or spark the creation of a standalone company. Or you may realize that your solution works best as a feature within someone else's platform, and begin exploring partnerships or licensing models instead.

The most important thing early on is to know what success would look like for *this phase* of the journey. Your goal right now is not to map out your ten-year roadmap. It is to clarify whether you're solving a product-level, venture-level, or system-level problem, and to define success in terms of learning, validation, and early traction.

CONSIDER THESE QUESTIONS

As you reflect on your own project or idea, consider the following:

- Are you building something that requires its own infrastructure, or could it live within an existing system?

- Do you want to be the person leading a company, or would you prefer to design something that someone else scales?

- Are you solving a narrow, defined pain point, or addressing a more structural issue in how care, diagnostics, or data flow?

- Will the success of your idea depend on reimbursement, adoption by hospitals, or regulatory clearance?

- Is your best pathway through product design, operational redesign, or strategic influence?

None of these answers are better than the others. But your clarity here will shape how you move forward, who you talk to, what kind of evidence you need to generate, and where you ultimately want your idea to land.

SUCCESS CAN LOOK LIKE MANY THINGS

This book will guide you through frameworks that are useful whether you are building a startup, spinning out a venture, or designing a new system from the inside.

Some of you will go on to raise money, build teams, and launch companies. Others will realize your solution is better suited for partnership or licensing. Still others will become invaluable intrapreneurs: people who drive innovation from within existing institutions.

Success might be a company. It might be a contract. It might be a pilot inside a clinic. It might be an open-source tool that other builders adopt and improve. What matters is not what the outside world calls it. What matters is that it works, that it's grounded in real need, and that it improves lives.

The next chapter will help you start identifying those needs. But for now, know this: what you're building doesn't have to fit someone else's mold. It just has to be real, and it has to move.

CHALLENGE: Exploring Your Builder Identity

Goal: Begin building your self-awareness as a future innovator by mapping out what kinds of building styles and success stories resonate with you. You're not picking your path yet; you're getting to know the terrain.

Instructions: Answer each prompt with honesty, not ambition. These questions are about you, not your idea. By the end, you'll have a personal lens through which to filter future opportunities, collaborators, and decisions.

Part 1: Builder Resonance Quiz

Read the following brief scenarios and note which one *feels* most energizing to you, not what you think sounds most impressive.

A. You're running a team of five, iterating on a prototype in a shared lab, and pitching for non-dilutive grants.

B. You're working nights to redesign how patient intake forms are handled in a local clinic. It's not "yours," but your ideas are making it smoother.

C. You're collaborating with a professor and a larger company to help license a new wound-healing adhesive you helped validate in a class project.

Which one sounds most like you? Why? (Write 3–4 sentences.)

What Your Answer Might Suggest

- **If you picked A:** You may be drawn to *venture building from scratch*. You thrive on ownership, iteration, and testing ideas in the lab or studio. This often aligns with roles as a founder or technical lead where you are creating something novel and guiding it toward company formation.

- **If you picked B:** You may be energized by intrapreneurship or systems-level improvement. You notice inefficiencies others accept as normal and find satisfaction in solving them. You might excel in startups focused on workflow, service, or software changes where impact comes from rethinking processes rather than inventing a new technology.

- **If you picked C:** You may be motivated by translation and partnership. You like working at the intersection of academia, industry, and business to move promising technologies out of research settings and into the market. This aligns with roles in tech transfer, licensing, or co-founding ventures around existing IP.

Part 2: Defining Early Impact

Write 2–3 examples of what "impact" could look like to you in the next 6–12 months (*even if you don't have an idea yet*).

Some examples to borrow or adapt:

- Hosting an interview with a hospital admin that sparks a real insight

- Getting invited to a team working on a clinical problem you care about

- Creating a new intake protocol that a local clinic tests

- Seeing your early sketch make someone's life easier, even once

This list should not be performative; it should be practical. What do you want to do, learn, or experience that would signal forward motion?

Part 3: Your Preferences and Thresholds

Answer the following in a few sentences each:

1. Do you imagine yourself leading something? Or are you more excited about building something others can lead?

2. How much ambiguity do you tolerate? Be honest. Do you thrive with open-ended projects or prefer defined roles?

3. What are two types of projects you're most likely to stay motivated on? (e.g., patient-facing, data-heavy, logistical, creative, clinical-adjacent, etc.)

4. What would make this journey not worth it for you? (This question matters later when we set up your personal "kill criteria.")

Anchor This

You'll revisit this builder profile later, once you've discovered a need and explored potential solutions. For now, save your answers somewhere you can find them again. They'll help you stay grounded in your why, even when your what changes.

CHAPTER 4: BUILDING YOUR NETWORK WITH INTENTION

When you're just starting out, networking can feel like a scary, daunting concept. Many universities will mention that you *should* network, yet few rarely teach you how to do it: let alone effectively. It's tempting to think networking means showing up at events, taking some pictures, collecting business cards, and hoping something sticks. But real networking begins before you shake a single hand.

PART 1: IDENTIFY WHO YOU NEED IN YOUR CORNER

The first step is knowing who it would actually be helpful to know. This sounds obvious, but most students skip it. They spread themselves too thin instead of mapping the specific categories of people who could shape the trajectory of their spinout.

Start by asking yourself: *Who are the people I will eventually need in my corner?* The list will look a little different for every founder, but for early healthcare and biotech innovators, a few groups almost always matter:

- **Venture Capital Firms:** If you live in or near a major city, there are likely dozens of VC firms with very different investment priorities. Don't try to know them all. Instead, focus on firms who invest in your stage and your space. If you're working on a medical device, that means firms with a track record of investing in devices (not fintech or consumer apps). Narrowing the field makes it possible to actually build relationships rather than chasing a blur of logos.

- **Specialized Lawyers:** Every spinout eventually runs into regulatory and intellectual property questions. Patent lawyers, FDA regulatory experts, and corporate attorneys aren't only service providers; they are often gatekeepers of introductions and industry knowledge. Even if you don't need them tomorrow, it pays to be on a first-name basis, or at the very least, know who you'd reach out to. You don't want to be scrambling for legal help when the clock is already running.

- **Other Founders in Your Space:** Perhaps the most overlooked category. You may initially want to avoid talking to other healthcare and biotech founders if you see them as competition. However, you're likely going to see the same people over and over at events, especially if you choose to stay in your college town after you graduate. These founders who are further along in their path can be helpful for sharing what pitfalls they've faced, who they've trusted, and what shortcuts they've discovered. Their experience is a playbook you can't Google. In addition, many of them will already know the types of people you are hoping to connect with. Getting at least one or two people on your side will help you build in the long run.

- **Suppliers and Manufacturers:** These connections are harder to come by at cocktail hours, but essential. Knowing who can actually make your device or source critical materials will be a bottleneck or a breakthrough. Keep an eye out for them at industry conferences or through introductions. You will likely not live in the same area as who you ultimately partner with in this area, but they tend to know who is doing what in their field and can redirect you if they can't manufacture something themselves.

- **Event and Program Managers:** These are the unsung heroes of an ecosystem. The person who runs a chamber of commerce innovation series or manages programming at a biotech hub often has more influence than the people on stage. They control the microphone: literally. When you attend one of their events, go find them and introduce yourself. It doesn't have to be long, but it takes maybe two minutes for you to share who you are, what you are building, thank them for their event, and ask them who they think you should talk to. Down the road, they may be the ones who put you on panels, amplify your news, or open doors you didn't even know existed.

A cardinal rule to remember is that every person you meet is more than a single contact. One good relationship multiplies into dozens, because you gain not just their support, but their entire network. That exponential effect is what separates the founders who stumble into opportunities from those who design them.

And here's where you, as a student, have a unique advantage. Many professionals assume you won't reach out. The bar for engagement is low. A simple expression of interest, paired with a thoughtful question or a short story about what you're working on, will land further than you think. In fact, it will often surprise people into helping you more than they might help a peer. Use that to your advantage.

Reflection – Map Your Network

Take 30 minutes to write down names (or at least target categories) for each of the following. Don't worry if you don't have names yet. Just knowing the categories you need to fill will guide your efforts in Part 2.

Venture Capital Firms

- Which firms in my city invest in healthcare/biotech?

- Which partners lead deals in my space?

Specialized Lawyers

- Do I know anyone in patent, regulatory, or corporate law?

- If not, who are the top firms in my area?

Other Founders

- Who are 3-5 local healthcare/biotech founders working in similar areas?

- Are there other student or alumni founders I can reach out to?

Suppliers and Manufacturers

- What kind of supplier/manufacturer would I eventually need?

- Do I know someone who might introduce me to them?

Event and Program Managers

- Which organizations consistently run events or publish content I can learn from?

- Who are the communications or programming leads behind those efforts?

Reflection Prompt:
Looking ahead over the next one to two semesters:

- Which of these categories feels most valuable for where I am right now?

- Which will matter most for where I want to be in a year?

- Who do I want to learn from directly?

- What am I struggling with and who could help me?

Write down one area that feels most critical. That's where your first networking effort should go.

PART 2: SCOUT EVENTS AND PLUG INTO NEWSLETTERS

Once you've mapped who you need to know, the next step is figuring out where those people actually are. You don't need to attend every networking opportunity that pops up. Instead, place yourself in the flow of information and gatherings that matter most to your spinout.

Where to Look for Events

Local Startup & Innovation Gatherings

- **Chamber of Commerce events** often include innovation or healthcare-focused programming. These are a great way to meet both founders and the program managers shaping your local ecosystem.

- **University demo days or pitch nights** bring together local investors, alumni founders, and service providers looking for the next wave of companies.

- **Healthcare and biotech meetups** (often hosted by accelerators, medical societies, or incubators) put you in the same room with founders who've already navigated clinical or regulatory hurdles.

Industry Conferences & Trade Shows

Larger events can be intimidating, but they're where you'll often find the suppliers, manufacturers, and regulatory experts who are harder to meet casually. You don't need to attend every major conference. Start with the ones closest to your stage or region and target events that are free for students or have discounted pricing.

A global anchor worth knowing:

- **BIO International Convention**: the largest biotech gathering in the world, held annually. It convenes investors, founders, suppliers, service providers, and policymakers. Even if you can't attend every year, following their programming and content keeps you aligned with industry trends.

State and Regional Organizations

Most states have a life sciences or innovation organization that acts as a hub:

- **SCBio** (South Carolina)

- **MassBio** (Massachusetts)

- **NC Life Sciences Organization (NCBIO)**

- **Ohio Innovation Exchange**

These groups often run pitch events, founder programs, and industry meetups that bring the right mix of investors, service providers, and founders into one room.

A quick Google search for "CITY innovation organization" or "STATE biotech organization" usually reveals the main players.

Once you find them, subscribe to their newsletters and follow their LinkedIn/X accounts.

Community-Driven Nights

Even general startup demo nights or tech mixers can be useful. While they may not always be biotech-heavy, they help you practice telling your story and widen your network outside of your immediate niche. Sometimes the most valuable connections come from unexpected places.

Newsletters, Substacks, and Digital Channels

Not all networking requires showing up in person. Many opportunities flow through digital channels. Start by subscribing to:

- **State/Regional Orgs**: Most (like SCBio or MassBio) publish event calendars, funding opportunities, and founder spotlights.

- **BIO International Newsletter**: Industry-wide news, policy updates, and global event listings.

- **Health Tech & Biotech Substacks**:
    - *The Health Tech Stack* (health innovation insights)
    - *Biotech Primer* (digestible biotech updates)
    - *Andreessen Horowitz Bio + Health* (VC perspective on biotech trends)

- **LinkedIn Thought Leaders**: Follow local founders, investors, and innovation hubs. Their posts often reveal opportunities faster than formal channels.

The simplest way to stay in the flow: pick **two events and two newsletters** this semester. Anything more than that will overwhelm you.

By scouting events and plugging into communication channels, you stop relying on chance to bump into the right people. Instead, you build a system for showing up in the rooms, emails, and conversations where opportunity lives.

This doesn't mean you have to attend something every week. It means you know where to go when you *do* want to invest your time. Networking is a marathon, not a sprint, and your calendar should reflect that.

Reflection – Event & Information Sources

Events

- Which recurring local events consistently bring together the people on my Step 1 list?

- Are there student-discounted or free events I can target this semester?

Newsletters/Groups

- Which 2-3 newsletters should I subscribe to that will keep me updated on biotech/healthcare opportunities?

- Are there state or regional biotech organizations I should follow?

- Are there Substacks or creators on LinkedIn, X, YouTube, and TikTok I can follow for ongoing insights?

Reflection Prompt
If I could only pick two recurring events and two newsletters to follow this semester, which would give me the highest leverage for meeting the people I identified in Step 1?

PART 3: Set Networking Goals

Once you know who you want to meet and where to find them, the next step is making sure your time at events actually counts. Walking into a room without a goal often means walking out without progress. Instead, treat networking the same way you'd treat an experiment: set measurable goals, track them, and reflect on what worked.

Networking can feel random. Some nights you meet no one relevant, and other nights you walk away with five new introductions. Goals help you cut through that uncertainty. They give you something to aim for, keep you accountable, and prevent you from drifting toward people you already know.

How to Set Practical Goals

Set a Minimum Contact Number

At each event, decide how many new people you want to meet. In general, a good baseline:

- **3 new contacts per event.**

That number is high enough to push you outside your comfort zone, but realistic enough to be sustainable. Start with 1 if you are more introverted though and just starting out.

Diversify Who You Talk To

Don't let all your conversations come from one category.

⊠ **For example:**

- 1 investor
- 1 founder
- 1 service provider (lawyer, program manager, etc.)

This ensures your network grows in a balanced way instead of clustering in one circle.

Track Your Progress

Keep a simple log after each event:

- Who did I meet?
- What category do they fall into (VC, founder, lawyer, etc.)?
- Did I learn something new?
- Is there a follow-up action?

This takes five minutes but creates a long-term record of your networking growth.

Tips for Making Networking Goals Stick

- **Arrive Early:** It's easier to meet people when the room isn't full and conversations haven't settled into groups.
- **Don't Linger Too Long:** If you've had a good conversation, exchange info and move on. The goal isn't depth on the spot. It's creating a reason to follow up later.

- **Push Yourself Toward Strangers:** Monitor yourself during the event. If you realize you've only spoken with people you already know, reset and approach someone new. If you can do it naturally, position yourself near the refreshment area or the restrooms to catch people as they're coming out, when they're alone and resetting, makes it easier to start a conversation. If you try the restroom strategy though, give enough space so it feels like a natural run-in rather than hovering too close. It should feel casual, not forced.

Preparation and Reflection – Event Goals

Before an Event

- How many new contacts do I want to make tonight?

- Are there specific categories I'd like to prioritize (VCs, founders, program managers)?

After an Event

- How many people did I meet?

- Who do I need to follow up with?

- Did I learn something useful about funding, scaling, or publicity?

Reflection Prompt

Which part of networking feels hardest for me? Starting conversations, diversifying who I talk to, or following up? What's one small tweak I can make at the *next* event to improve?

PART 4: DO THE NETWORKING

You've mapped who you need to know (Step 1), scouted events (Step 2), and set goals for yourself (Step 3). Now comes the part that feels most intimidating: actually walking up to people and talking to them. As a student, you'll often find yourself in rooms with professionals who are 20+ years older and far more established. The good news: they *expect* students to be curious and learning, not polished experts. If you approach conversations with preparation and genuine interest, you'll stand out.

Start With Who You Know

If you arrive at an event and feel overwhelmed, start by finding someone you already know, like a professor, a peer, or another founder.

- **Use them as a bridge.** Ask, "Who else do you think I should meet tonight?"

- Let them make an introduction on your behalf, which is often easier than approaching on your own.

Connect With the Hosts

As mentioned before, the people running the event (program managers, organizers, marketing staff) are some of the most valuable connections in the room.

- Introduce yourself early: "Hi, I'm [Name], I'm a student working on [brief project/area]. I really appreciate you putting this together and wanted to come personally thank you. Do you have any other events like this going on in the near future?"

- Before you end the conversation: "Is there anyone here tonight you think I should meet?"
  Organizers often know everyone in the room and are happy to guide students to the right conversations.

Introduce Yourself Clearly and Briefly

Professionals don't expect students to have a perfect pitch, but they do expect clarity. Aim for a **30-second introduction**:

- Who you are (student, program, or lab).

- What you're working on (one sentence).

- Why you're there (learning, exploring, building).

⊗ Example:
"Hi, I'm Priya. I'm a bioengineering student at Stanford working on a wearable for stroke rehab. I'm here to learn more about how startups in this space think about funding and regulation."

That's enough to spark curiosity without overselling.

Make the Conversation Meaningful for Them Too

Networking shouldn't feel one-sided. If you only take, the professional will feel used. Instead:

- Show curiosity about their work: "That project sounds fascinating. What impact are you most excited about?"

- Offer something back: "I saw an article about [topic they mentioned]. Would you like me to send it to you?"

- Listen more than you speak. Professionals remember when students make them feel valued, not drained.

Ask Meaningful Questions

Asking thoughtful, open-ended questions gives you insight and shows respect:

- "What drew you into this field?"

- "If you were starting out as a student today, what would you focus on?"

- "What trends in biotech/healthcare are you most excited about right now?"

- "What's one thing you wish founders did better when they approached you?"

How to Approach Small Groups

Walking up to a group that's already deep in conversation can feel intimidating. Use these tactics:

- **Pick groups of two or three, not six.** Smaller circles are easier to join.

- **Wait for a natural pause.** Make eye contact and smile before speaking.

- **Introduce yourself simply.** "Hi, I'm [Name], I hope I'm not interrupting. Mind if I join you?" Most groups will open the circle and include you.

- If the topic is technical and you're lost, lean into curiosity: "I'm new to this space, could you explain what that means?" Most professionals will be flattered to teach.

Exit Gracefully

Conversations don't need to be long to be effective. When you wrap up:

- **Echo their name before leaving.** "Great meeting you, Dr. Lee! I'll send you that article we talked about." Repeating their name reinforces memory and makes them feel truly seen.

- **Signal next steps.** "Can I connect with you on LinkedIn?" or "Thanks for sharing that insight, I'd love to follow up later."

This shows respect for their time and creates a natural transition.

Preparation – In-the-Room Networking

Before the Event

- Who do I already know that I can use as a bridge?

- Who is hosting the event, and can I introduce myself to them early?

During the Event

Did I give a clear 30-second intro each time?

- Did I ask at least one meaningful question per conversation?

- Did I echo their name before walking away?

- Did I talk to at least one person I didn't already know?

Reflection Prompt

When I think back on events I've attended, did the conversations feel meaningful for both me *and* the professional, or only for me? How can I balance curiosity with giving value in my next interaction?

PART 5: FOLLOW UP STRATEGICALLY

Meeting someone at an event is just the first step. Real networking happens in the follow-up. A weak or nonexistent follow-up leaves your new contacts as strangers; a thoughtful one turns them into potential collaborators, mentors, or champions.

Why Follow-Up Matters

People meet dozens of faces at each event. If you don't follow up, you'll be forgotten. Following up is how you:

- **Anchor the connection**: Remind them who you are and what you talked about.

- **Show genuine interest**: Demonstrate you care about *their* work, not just your own.

- **Open the door**: Create a reason for ongoing interaction.

How to Follow Up Well

Send a LinkedIn Connection

Within 24-48 hours, send a LinkedIn connection with a personalized, short note. Include:

- **Where you met** ("Great meeting you at [Event Name]").

- **Something specific you discussed** (an anecdote or shared interest).

- **A light touch next step** (e.g., "Looking forward to staying in touch").

Once this is done, if you need to reach out to them in the future, they will have this message as a refresher for where and how they met you. All in all, it will make it easier to re-engage later.

Swap Business Cards (Digital Preferred)

Always have a business card ready for people you want to build a relationship with outside a casual acquaintance level. Even better, use a free digital card app like **Blinq**. It instantly saves your info to their phone, reducing the chance they lose track of you.

Capture Their Interests

Don't leave a conversation without knowing what they're interested in. Are they raising a fund? Hiring talent? Exploring a new research area? This gives you something to send them later, whether it be a relevant news article, an introduction, or even just a note of encouragement.

Follow Up with Value

Instead of a generic "great to meet you," try sending something useful:

- A news article tied to their interest.

- A follow-up question about something they mentioned.

- A relevant event they might want to attend.

Even small gestures stand out because they show you were listening.

What Not To Do

- **Don't oversell yourself.** Your first follow-up should not be a pitch deck or a funding ask.

- **Don't spam.** If you don't hear back right away, wait until you have a real reason to reach out again.

- **Don't be transactional.** Focus on building a relationship, not extracting favors.

Reflection – Strategic Follow-Up

Immediately After an Event

- Did I connect with each new contact on LinkedIn?

- Did I include a personal detail or anecdote in the note?

- Did I capture what they're interested in for future reference?

Within One Week

- Did I send something of value (article, event, intro) to at least one person I met? (Best done for people you are trying to deepen a connection with. You can skip this part if they are best served as an acquaintance for now.)

- Did I log the important people of my new contacts somewhere I can track them?

Reflection Prompt

When I think back on the last few networking interactions I had, did I follow up in a way that added value? Or did I just let the conversation fade? What's one improvement I can make in my next follow-up?

PART 6: SET YOURSELF UP FOR SUCCESS

Networking doesn't stop at events. The easiest way to keep people in your orbit is by reminding them what you're working on. Social media, especially LinkedIn, is a low-effort way to do this.

Why LinkedIn Isn't Going to Make You Uncool

I know. Posting on LinkedIn can feel cringe and the last thing you want is to end up on the LinkedInLunatics subreddit. However, it does serve a purpose. Most people won't think to reach out unless they see you doing something relevant. By posting updates, you:

- **Stay visible**: Your name stays fresh in their minds.

- **Signal progress**: People want to support founders who are moving forward.

- **Invite opportunities**: Someone might connect you with a resource simply because they saw your post.

How to Do It Without Being "Cringe"

- **Post once a month.** That's enough to stay active without overwhelming yourself.

- **Share progress, not polish.** People love to see learning in action.

- **Keep it simple:**
    - An update on your project: "Excited to test our prototype next week."
    - Gratitude: "Thanks to [org] for hosting an incredible demo night; I learned so much."
    - Reflection: "One lesson I've learned building [project] is..."

Reflection Prompt

If someone from an event checked my LinkedIn today, would they see what I'm working on and where I'm headed? Or would they see silence?

Anchor This

Most students treat networking as a side activity, like something you "should" do, but not central to building your company. The reality is the opposite. The strength of your network will often determine how quickly you can raise funding, find a regulatory expert, land your first pilot, or get invited onto a stage where investors are listening. The earlier you treat networking as part of your work, not an afterthought, the more leverage you'll have when opportunities arise.

As a student, you have a unique advantage: expectations are low. People assume you're still figuring things out, which means even a small effort, such as — introducing yourself clearly, asking a thoughtful question, or following up with a useful article — lands far harder than it would for someone more established. That window doesn't last forever. Use it.

Networking isn't about collecting contacts. It's about creating systems that keep you in the flow of opportunities:

- You mapped who you need to know.

- You found where those people gather.

- You set goals that keep you accountable.

- You learned how to walk into a room and hold your own.

- You followed up and added value.

- And you maintained visibility so people know where to plug in.

The truth is, you don't need hundreds of connections. You need a dozen or two who trust you, remember you, and are willing to open doors when you ask. If you approach networking with that mindset, you'll be far ahead of peers who only think about it when they're desperate for funding or a job.

So here's your playbook: **show up prepared, ask good questions, follow up with intention, and stay visible.** If you do those things consistently, networking stops being awkward and starts becoming one of the most powerful levers you have for turning your idea into a real company.

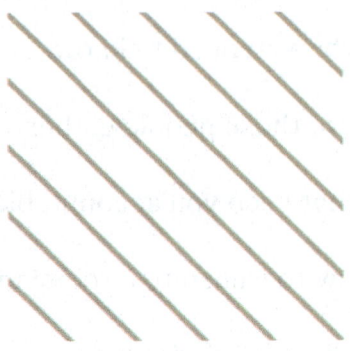

Part II

Problem Discovery

CHAPTER 5: WHERE TO FIND REAL PROBLEMS

Everyday immersion, clinical, academic, and personal sourcing strategies

Every innovation starts with a problem. But not every problem is worth solving, and not every path to identifying one is intuitive: especially when you're just starting out.

Students, early-career professionals, and aspiring innovators often make the same mistake: jumping into solution mode without first grounding themselves in a real-world need. The result is predictable. They build something clever, technically sound, and beautifully designed that never actually gets adopted, because it solves a problem no one was really struggling with.

If the previous chapter helped you understand what kind of innovation you might be building (product, venture, or system), this chapter helps you find the real-world problems worth building *for*.

At BMTT, we don't believe in waiting for perfect access or polished settings to start this process. Our needs sourcing strategy is built to be inclusive, realistic, and actionable. Whether you're a student at a university with no hospital in sight, or you're embedded in a clinical environment but unsure what to look for, this chapter will show you how to begin.

Start With This: You're Not Hunting for Ideas. You're Listening for Friction.

Finding a "real problem" doesn't mean waiting for someone to say, "This is broken. Can someone invent something better?" It means looking for friction. That friction could be inefficiency, delay, confusion, danger, redundancy, emotional distress, or cost: anything that suggests the current state of things is falling short.

One of the first times I ever encountered this kind of friction, I almost didn't recognize it for what it was. I was shadowing a cardiologist in high school when, after completing a routine angioplasty, he casually said, "Well, I'll be seeing them again in less than two years."

At the time, I didn't know enough to question the comment directly, but it stuck with me. Was that frustration patient-specific? Disease-specific? Systemic? Procedural? As it turns out, the answer was all of the above.

That passing remark became the first real clinical need I ever chased. What I eventually learned was that angioplasty wasn't actually removing plaque from arteries. It was simply compressing it into the arterial wall to create temporary space, followed by the placement of a stent. These stents often carry some of the highest mortality rates associated with hospital-acquired infections. They are designed to work only if restenosis occurs (meaning they rely on the artery narrowing again just to stay in place). At the end of the day, they serve as additional points of plaque accumulation.

The physician wasn't necessarily criticizing the procedure. But something in the tone of his voice (routine and resigned) made me start asking questions. That moment became the foundation of my first startup, and ultimately, the reason I do the work I do today.

That's the mindset you have to develop. Not everything broken will look broken. Not every comment will come with urgency or outrage. Sometimes, the clue is in what people accept as normal. You have to train yourself to question everything.

That friction lives in many places. Some of them are obvious, like a surgical suite. Others are subtle, like a routine phone call between a nurse and a caregiver.

This chapter walks through five structured domains you can explore for problem sourcing:

1. Everyday immersion and personal networks

2. Clinical observation and shadowing

3. Academic procedures and literature

4. Legal and litigation datasets

5. BMTT's idea intake and bootstrapped alternatives

1. EVERYDAY IMMERSION: FINDING PROBLEMS IN YOUR PERSONAL NETWORK

The easiest and most overlooked source of clinical insight is the world around you. Real needs are visible if you ask about them. Ask your friends, family members, and acquaintances about their experiences navigating the healthcare system. Ask what was confusing, frustrating, delayed, inefficient, or emotionally exhausting. Ask what happened when a device didn't work the way it was supposed to, when a doctor couldn't reach them, or when instructions didn't make sense.

A commonly retold segment from Ernesto Sirolli's TED Talk: *"Want to Help Someone? Shut Up and Listen"*, tells a story about planting Italian tomatoes in Zambia, only to watch hippos eat the entire crop. The locals shrugged and said, "Of course, we knew that would happen." The lesson was simple: you can miss the most obvious barrier if you don't ask the people living with the problem. The same is true in healthcare. Listening deeply, even before you start building, is the fastest way to uncover problems that actually matter.

As innovators, we have to check our confidence. Many gaps are gaps for a reason. Sometimes a process looks inefficient from the outside, but the people inside it know about hidden constraints that make it hard to change. Oversimplifying what looks like an "easy fix" can frustrate the very people you want to help. Patients, caregivers, and clinicians often have already tried the obvious solutions and know why they failed. The only way to learn those lessons without wasting time or breaking trust is to ask, listen, and design alongside them.

You're not collecting solutions. You're collecting signals... patterns of frustration, friction, or failure.

Who to Ask:

- Patients with chronic illnesses or complex medication regimens

- Parents or caregivers

- Nurses, pharmacists, techs, EMTs, and administrative staff

- Family physicians or outpatient care teams

What to Ask:

- What part of that process didn't work?

- Where did you feel like you were waiting too long?

- What's something you have to do every day that you wish was simpler?

Often, it's the most "boring" and repeated problems that signal scalable opportunities.

✳ **BMTT Tip:** Don't limit your questions to just patients. Ask healthcare workers what repeatedly slows them down. Workflow inefficiencies are often where the best ideas hide.

2. SHADOWING AND CLINICAL OBSERVATION: FRICTION IN ACTION

If you have access to hospitals, clinics, or even outpatient care, shadowing is one of the most powerful ways to source clinical needs. This is where you stop theorizing and start watching.

Shadowing allows you to observe the hidden complexities of real care environments. You're not there to solve anything. You're there to understand what happens when things break down (or nearly do).

How to Observe:

- Be unobtrusive. Watch, don't disrupt.

- Focus on workflow, not just technology.

- Watch for workarounds. Every time a nurse says, "We usually do it this way, but today I had to…" that's a signal.

- Note delays, dropped tools, repeated explanations, moments of patient confusion, and every time a clinician has to choose between speed and accuracy.

How to Ask Questions:

- Wait for calm moments, not high-stakes procedures.

- Ask open-ended questions like: "What's the most frustrating part of this process?"

"Is there anything here that slows you down unnecessarily?"

"If you could change one thing about how this is done, what would it be?"

What to Document:

- Date, time, and setting

- What exactly was happening

- What tools or technology were being used

- Body language, tone, delays, breakdowns, or failures

- Insider language or slang (these often point to cultural workarounds)

Even if you only get a few hours of observation, detailed field notes can yield weeks of insight.

3. ACADEMIC LITERATURE AND PROCEDURAL STUDY: PROBLEMS HIDDEN IN PLAIN SIGHT

When clinical access isn't possible, deep research is your best friend. BMTT and Stanford Biodesign both emphasize the importance of academic grounding in your observation process.

Where to Look:

- **PubMed and Google Scholar:** Find systematic reviews, device trials, and clinical outcome studies

- **ClinicalTrials.gov:** Spot where new interventions are being tested, and what gaps they aim to fill

- **Surgical and procedural manuals:** Study protocols and identify bottlenecks or overly complex steps

- **UpToDate and eMedicine:** Look for consensus guidelines that point to areas of uncertainty or risk

What to Look For:

- Repeated adverse events or device failure modes

- Comparisons between procedures or technologies

- Commentary on cost-effectiveness, safety, or patient satisfaction

- Notes on "limitations" in discussion sections: they often hint at known gaps

※ **BMTT Tip:** Once you identify a frustrating or inefficient procedure, talk to clinicians about it. Ask what part of the protocol is outdated, skipped, or routinely patched with workarounds.

4. LEGAL DATA AND LITIGATION: LEARNING FROM WHAT WENT WRONG

One of the most underutilized sources of clinical insight is the legal system.

Every time a lawsuit is filed over a faulty device, delayed treatment, or systemic failure, a problem is being documented in full detail. BMTT teaches founders to use this data not to build defensively, but to identify gaps in care and design around failure points.

Where to Look (Free Resources):

- **Google Scholar (Case Law)**

- **Justia (Healthcare Litigation)**

- **FDA MAUDE Database:** Search device names and failure modes

- **State and Federal Court Websites:** Look for medical device or malpractice rulings

Where to Look (Paid Resources):

- **LexisNexis and Westlaw:** Ideal for deep litigation dives, especially for universities with subscriptions

What to Look For:

- Patterns in device failure or improper use

- Malpractice themes: missed diagnosis, improper documentation, post-op errors

- Class-action suits around poor software integration or drug delivery inconsistencies

These aren't just cautionary tales. They are validated records of problems that caused harm and demand better solutions.

※ **BMTT Tip:** You can now automate much of this monitoring with tools like ChatGPT. At BMTT, we've experimented with custom GPT agents configured to scan and summarize newly published lawsuits, medical device recalls, and FDA safety communications. These agents can run on a schedule and feed relevant data into digestible summaries or flag entries that match your area of interest. For independent builders without full legal database access, this kind of automation can be a game-changer by allowing you to stay on top of signals without manually combing through dense databases every week.

5. BMTT INTAKE, BOOTSTRAPPING, AND ALTERNATE ACCESS POINTS

Not everyone has access to hospitals, labs, or legal databases. That doesn't mean you're shut out. In fact, some of the most insightful student teams we've worked with at BMTT have sourced their best ideas by leaning into resourcefulness, not credentials.

We've built our own infrastructure to support this at scale. For example, BMTT operates a public-facing **clinical need intake portal** that allows healthcare professionals, patients, and technologists to submit real problems anonymously. Those submissions are then triaged and organized by custom GPT tools we've developed internally, allowing us to spot patterns across entries and flag high-potential problems for follow-up.

But this kind of infrastructure isn't just for institutions. Students have replicated similar systems on a smaller scale to great effect. We've seen teams create simple intake forms using tools like Google Forms or Typeform, paired with email or social media distribution strategies. By sending these forms to friends, family members, clinicians in their network, or even university listservs, they're able to collect dozens (sometimes hundreds) of real, lived problems that become the basis for serious venture ideas.

※ **BMTT Tip:** When building your own intake form, focus on clarity. Ask for one problem per entry. Make it anonymous if possible. And keep your questions focused on experience, not opinions. Instead of asking "What do you think we should build?"

ask "What part of your workflow or care experience is most frustrating, confusing, or broken?"

Other Bootstrapped Observation Strategies:

- **Reddit:** Subreddits like r/medicine, r/AskDocs, r/AskNurses, and r/medicalschool are filled with raw, unfiltered frustrations and everyday workarounds.

- **YouTube and TikTok:** Watch actual clinicians or patients describe what they do. You'll spot repeated patterns of friction that are rarely mentioned in literature.

- **Medical Supply Catalogs:** Browse for products used in routine care and then search their recall history, device complaints, or failure rates online.

- **Hospital Job Descriptions:** These often reveal hidden inefficiencies or overload; look for descriptions that suggest administrative burden, workflow bottlenecks, or underutilized expertise.

- **Google Alerts:** Set up alerts with keywords like "[device name] malfunction," "[procedure] complications," or "[treatment] side effects." You'll start to pick up on what's failing in the field.

- **Scheduled GPT Monitoring:** You can also configure custom GPTs or similar tools to run scheduled searches across public data sources (news releases, case law summaries, regulatory updates, and recall logs) and then return relevant findings. While not a replacement for deep research, these systems act as passive sentries, surfacing problem signals while you focus elsewhere.

You do not need a badge to find problems worth solving. What you need is consistency, creativity, and a system for listening well. If you approach need-finding with discipline, even without clinical access, you will find opportunities that others miss.

Final Note: Don't Rush the Problem Framing

You will be tempted to solve early. Don't.

One of the most damaging habits we see is early-stage founders choosing a problem, sketching a solution, and skipping past the hard work of validation and problem-framing. Every time you catch yourself saying "what if we just..." take a breath and ask: **"Do I understand the problem well enough to build anything yet?"**

Your job in this phase is not to create. It is to observe, listen, and document. The better your problem sourcing, the stronger your solution's foundation will be...

CHALLENGE: Your First 20 Problems
(Skip if You Have Your Problem)

Your goal for this challenge is simple in structure but powerful in outcome: identify and record at least 20 real-world problems that reflect unmet needs, broken workflows, or inefficient processes.

Why twenty? Because innovation is a funnel. Most problems will not survive closer inspection. Some will be too narrow, some too broad, some already solved, and some outside your reach. If you only write down three or four, the odds that one is worth pursuing are slim. By forcing yourself to gather a larger set, you create room for filtering and refinement: exactly how professional innovation programs operate.

This number is also about practice, not perfection. Spotting problems is a skill. At first, you may only notice obvious frustrations, but as you push toward twenty, you'll begin to see subtler friction points and patterns that you might have overlooked. This is how Biodesign teams, and other structured innovation sprints, train students to see opportunities more sharply. The exercise itself builds your "problem-finding muscle."

For now, you are not solving anything. Your job is only to document friction. Each problem is a potential entry point into something bigger, and the more entry points you generate, the more likely you are to find one worth walking through.

Instructions

Gather from multiple domains.
Don't rely on a single source. Use at least **3 different sourcing strategies** from the chapter (e.g., personal conversations, Reddit, shadowing, medical literature, supply catalogs, legal case review).

1. **Document the friction clearly.**
   Use a simple structure:

   - **Where did you find it?** (e.g., Reddit, grandparent's clinic visit, clinical article)

   - **What was the friction?** (What happened or failed to happen?)

   - **Why might it matter?** (Don't validate: just guess at impact if unresolved.)

2. **Don't filter too early.**
   Even if the problem seems small or "not worth solving," write it down. You're casting a wide net. High-impact insights often hide behind boring complaints.

3. **Push past the obvious.**
   The first 5–7 problems might come easily. But the later ones, the ones that make you observe more deeply or reframe what you thought you knew, are usually the most interesting.

Optional Worksheet Format (You can replicate this in a notebook, Word doc, or spreadsheet)

Problem #	Source	Friction Point	What Happened / Didn't Happen
1	Reddit (r/AskNurses)	Medication handoffs	A nurse reported patients often get conflicting medication lists during transfers.
2	Family caregiver	DME setup confusion	Grandma got a wheelchair but didn't know how to lock the brakes. No training.
3	Clinical shadowing	Device swap delay	The surgical team paused for 3 minutes because they couldn't locate the right suction tip.

Tips to Expand Your List

- Revisit frustrations you've experienced firsthand in healthcare, school, or caregiving.

- Ask a friend who has a chronic illness what part of their care is annoying or unclear.

- Search for "I hate when..." in Reddit health-related threads and read the comment chains.

- Check the FDA MAUDE database for device complaints that seem to happen more than once.

- Browse recent malpractice summaries for common procedural errors or missed diagnoses.

- Skim 10–15 YouTube videos by nurses, patients, or physicians and write down any bottlenecks or tech failures they mention.

- Look through 10 hospital job listings and highlight repeated mentions of workflow, documentation, or supply-chain issues.

- Search TikTok hashtags like #nurselife, #caregiverlife, or #medschool and watch how real users talk about inefficiencies.

Target Output

- **Minimum: 20 raw problems** written clearly and concisely.

- **Stretch Goal: 30–35 problems** across at least 4 sources.

- Keep your notes. We'll return to these in the next chapter and begin refining them into structured need statements.

Anchor This

Don't fall in love with any one problem yet. The goal here isn't to identify "the best" need; it's to fill the funnel. You are learning to think like a builder: observant, skeptical, and open-minded. Your only job right now is to document the world as it is, so you can later imagine what it could be.

You're not just building a startup; you're building a lens. Keep looking. Keep writing. Let the data do the talking.

CHAPTER 6: WRITING THE RIGHT PROBLEM

High-Impact Need Statements, Framed to be Built on

The best founders do not fall in love with their solutions; they become obsessed with understanding their problems. If the last chapter taught you how to find friction in the world around you, this one will teach you how to frame that friction into something actionable, clear, and testable.

Welcome to need statement writing. But don't worry: you're not being asked to pick your "one big idea" just yet.

At this stage, you should still be holding on to a broad list of potential problems: many of them raw, messy, or overlapping. That's by design. This chapter will help you practice translating several of those friction points into proper need statements. You are not committing to one idea. You're sharpening your lens so that, in the next chapter, you'll be able to narrow your funnel intelligently.

Why We Take This Seriously

At BMTT, we have seen dozens of well-meaning teams skip or rush this step, and the result is almost always the same: they start building something impressive that no one actually needs.

Even worse, we've seen countless teams jump into problem-solving mode the moment a physician or clinician says, "I wish there were a tool that could do X." That kind of shortcut might feel like you're getting insider access, but it's often a trap.

Here's why: authority bias is real, and it can destroy early-stage validation.

Physicians, while invaluable collaborators, are not always reliable problem framers. We've seen students build entire pitch decks around a physician's complaint, only to learn that the tool already exists, was simply never adopted at their hospital, or that the physician was never properly trained on how to use the full functionality of an existing solution. In many cases, those same clinicians have no influence over procurement or product adoption, and thus fall behind in what's actually available or evolving.

One student I worked with was encouraged by a physician to pursue a tool for better visualization of the cervical ring in hysterectomies. It sounded like a distinct need, and the student began mapping out a project around it. Only later did they learn that several visualization tools already existed; they simply weren't in use at that hospital. The student walked away with an important lesson: physicians can spotlight real frustrations, but those frustrations are not always evidence of an unmet need. Careful validation with broader perspectives is essential.

So when someone in a white coat says, "I wish we had a better way to do this," your job is to smile, nod, and quietly go verify whether that problem is real, whether others experience it too, and whether there's more to the story than what you're hearing.

This chapter will help you do that by teaching you how to reduce what you've heard and observed into a precise, flexible, and evidence-informed need statement.

STEP 1: TRANSLATE OBSERVATIONS INTO ROOT PROBLEMS

Before you write a single word, you need to go back to your field notes and isolate the **causal** friction. You are looking for a specific failure, delay, cost, or confusion that leads to a worse outcome for a clearly defined population.

Ask yourself:

- What exactly is going wrong here?
- Who is affected?
- What undesirable outcome results from that failure?

This is the foundation of your need. It is not your interpretation of what they "should" be doing. It is your summary of what is *actually* happening and why it matters.

Now reduce that to its core components:

- **Problem** (the specific gap or inefficiency)
- **Population** (the affected users, patients, or stakeholders)
- **Outcome** (the metric, experience, or impact being compromised)

Once you've clearly defined those three pieces, you can begin crafting a need statement.

STEP 2: WRITE A SOLUTION-INDEPENDENT NEED STATEMENT

A strong need statement should describe the problem *without suggesting a solution*. This is harder than it sounds.

Here's the formula we recommend starting with:

A way to [solve the problem] in [target population] to [improve the outcome].

⊠ **Example (well-framed):**
A way to reduce medication administration errors in elderly inpatients to prevent adverse drug events.

Common Mistake:

A way to build a smart pill bottle that alerts nurses when medications are missed.

The second example embeds a solution before the problem is even fully understood. The first example, by contrast, leaves open many potential paths forward. That flexibility is what allows for creativity, iteration, and better user alignment later in the process.

If you find yourself using specific technologies, device names, or assumptions like "automated," "AI-powered," or "mobile," take a step back. That's a sign you're already in build mode before the problem has been fully validated.

The only time solution constraints may be acceptable is when pursuing a highly incremental improvement to an existing tool or device, and even then, the framing must be deliberate and well-justified.

STEP 3: CHALLENGE AND VALIDATE THE SCOPE

A well-written need statement is both specific and open-ended. You want it to be narrow enough to act on, but not so narrow that you prematurely limit your solution space or market size.

Common Mistake: Too Narrow

A way to reduce bleeding in femoral artery punctures in left-handed cardiologists at community hospitals.

This is too restrictive. The need might exist, but it applies to too few users to justify commercial development unless it's part of a broader problem.

Common Mistake: Too Broad

A way to improve surgery outcomes for all patients.

This is too vague to be useful. It doesn't help you target stakeholders, define success metrics, or prioritize development.

※ **BMTT Tip:** The sweet spot often lies in phrasing that identifies a repeatable, observable problem, with a clear population, and an outcome that can be meaningfully improved. Then test that scope with experts and peers: does it resonate across multiple environments, or is it only true in the place you happened to observe it?

STEP 4: NEVER TAKE ONE PERSON'S WORD FOR IT

This cannot be overstated: **no single clinician, patient, or stakeholder speaks for the system**.

We've seen founders burn months and thousands of dollars pursuing problems that a single person complained about. Later, they find out the "problem" was solved years ago at another hospital or that nobody else experiences it the same way.

Always verify what you hear.

- Look for **pattern recognition**: do others in the same role describe similar frustrations?

- Go **cross-institutional**: does this problem exist in different cities, hospitals, or systems?

- Use **literature and databases**: does the data support what you're being told?

Your goal is to understand whether the need is widespread and real, or isolated and anecdotal.

STEP 5: CATEGORIZE THE TYPE OF NEED

Once your need statement is written and validated, categorize it by type:

- **Incremental**: Improves an existing tool, workflow, or treatment path.

- **Blue Sky**: Represents a new way of thinking about the problem; no obvious precedent exists.

- **Mixed**: Involves a combination of existing and novel ideas.

This will help you frame how radical your eventual solution needs to be, what kind of partnerships you may need, and how high the bar is for evidence and adoption.

STEP 6: BEGIN BUILDING NEED CRITERIA

Before you brainstorm a single solution, begin defining what a successful solution *would need to do* in order to be viable.

Ask:

- Where, when, and by whom will this be used?

- Does it need to be disposable, reusable, digital, portable, or intuitive to non-specialists?

- What are the cost constraints, time limitations, or safety concerns?

- Are there reimbursement codes that would be needed to support adoption?

This criteria list will become your blueprint later when you evaluate potential ideas. For now, it will help you stay anchored in reality while still thinking creatively.

Before we wrap up, let's walk through what this actually looks like in practice. The following short narrative follows a fictional student team as they move from a vague field observation to a well-scoped need statement. You'll see how each step of this chapter (triad framing, scope refinement, validation, and early success criteria) plays out in real time. It's not a perfect journey, but that's the point. The process is messy, and the clarity comes *after* the work.

CASE REFLECTION: FROM MESSY TO MEANINGFUL

Let's follow a fictional student team (Maya, Chris, and Lena) as they work to sharpen a rough observation from their clinical shadowing. Like many early teams, they weren't short on insights; they were just short on structure.

Day 1: The Gut Feeling
During a shadowing session in the emergency department, Maya noticed something chaotic but familiar:

"When patients get transferred from another hospital, it's a mess trying to figure out what meds they're on. The nurse spent almost 30 minutes calling around and digging through paper notes."

They jotted it down in their shared doc as:

"Med list chaos during ER handoffs."

It felt important, but they weren't quite sure what to do with it yet.

Day 2: The Overreach
Back in their workspace, Chris took the first stab at a need statement:

A way to create a centralized database for medication records during patient transfers.

It sounded polished, maybe even pitch-worthy, but it jumped ahead. Lena pointed out that they were already suggesting a solution without having defined the core problem. And even worse, they hadn't really identified *who* was struggling or *what* the consequence was.

They hit delete.

Day 3: Getting to the Root
This time, they used the triad approach from their BMTT notes.

- **Problem**: Medication discrepancies during patient transfers

- **Population**: Emergency department nurses receiving external transfers

- **Outcome**: Delays in treatment and increased risk of adverse events

Now they tried again:

A way to ensure accurate medication reconciliation in transferred patients to reduce delays and prevent adverse drug events.

This version felt clearer. No embedded solution. Just the friction, the people affected, and the consequence. They shared it with two nurses and a pharmacist. The feedback?

"That's our life," one nurse said. "We spend so much time trying to fix that."

Encouraging, but one pharmacist noted that similar solutions were already being tested at other hospitals. Time to test the scope.

Day 4: Zooming Out
After reviewing literature and talking to peers at other schools, they realized the issue extended beyond just the ER. Post-op units, psych facilities, and primary care clinics were all reporting similar frustrations around care transitions.

They revised again:

A way to ensure accurate medication reconciliation during patient transitions to prevent adverse drug events.

It wasn't flashy, but it was grounded, flexible, and validated. They tagged it as an **incremental** need and listed out early success criteria:

- Works across different EHR systems

- Doesn't add time to discharge

- Includes caregiver/family verification

- Could be adopted without new billing codes

They weren't ready to build yet, but they were ready to choose *whether* this need was worth building around.

Takeaway
Clear need statements rarely show up in one shot. They take revisions, conversations, and uncomfortable honesty. You won't always land on the right version first, but if you stay curious and structured, you'll find the signal.

FINAL NOTE: RESPECT THE DISCIPLINE

Writing a good need statement is slow, sometimes frustrating work. But if done correctly, it can save you months of wasted effort and thousands of dollars in misaligned development. More importantly, it will keep your team focused on what really matters: delivering something that solves a real, validated problem for a real, defined user group.

You are not expected to land on just one statement today. In fact, your next chapter is designed to help you evaluate and reduce the full list of validated needs you've generated: from many, to one. Before you can choose what to build, you must first learn how to compare what's worth building.

So pause. Test. Reframe. Pressure test again.

Then, and only then, start to funnel.

CHALLENGE: Sharpening The Signal, Not The Scope

In the previous chapter you generated 20 friction points. Those entries are raw. Your goal now is not to pick a winner, but to **sharpen and filter** so you finish with **10** clear, solution-independent need statements ready for deeper comparison in the next challenge.

Step 1: Quick triage to 14

From your 20, cut **6** that clearly fail any one test:

- The problem is vague or not observable.
- The primary user is unclear.
- The consequence is trivial or unmeasurable. Keep **14**. Do not overthink this pass.

Step 2: Frame the Core Triad for all 14

For each of the 14, write three lines:

- **Problem:** What exactly goes wrong, observable in the real world.
- **Population:** Who is directly affected.
- **Outcome:** What happens if it is not fixed.

Examples

- Problem: Conflicting medication lists at discharge
  Population: Elderly inpatients with multiple comorbidities
  Outcome: Readmissions due to drug interactions

- Problem: Supply misplacement during procedures
  Population: OR nurses and scrub techs in community hospitals
- Outcome: Delays that extend anesthesia time and increase cost

- Problem: Poor training on durable medical equipment at home
  Population: New family caregivers after hospital discharge
  Outcome: Falls and preventable ED visits

Step 3: Draft solution-independent need statements for all 14

Use one template and keep solutions out:

- A way to [solve the problem] in [target population] to [improve the outcome].

Examples

✓ A way to ensure accurate medication reconciliation in elderly inpatients to prevent adverse drug events

✓ A way to keep OR supplies findable and ready for nurses during procedures to reduce avoidable delays

✗ A wearable that alerts doctors to drug conflicts in real time

Step 4: Scope check to 12

Read each need statement aloud and ask:

- Is the population specific enough to recruit in real life.

- Is the problem observable, not just a feeling.

- Is the outcome something you can measure or at least proxy.

Remove the 2 weakest. Keep 12.

Mini example of a fix

- Too broad: A way to improve communication in hospitals to reduce errors.

- Sharpened: A way for night-shift nurses in med-surg units to verify critical lab value handoffs to prevent missed interventions.

Step 5: Tag the type of need for all 12

Label each as **Incremental**, **Mixed**, or **Blue Sky**. This sets expectations for effort and risk later.

Step 6: Write initial success criteria for 6 items

Pick 6 of the 12 that feel strongest today. For each, list 5 to 7 design constraints that any viable solution must respect. You are not listing features. You are describing reality.

Examples

- Medication reconciliation need
  Constraints: Works within EHR or low-tech workflow, usable by nurses on discharge day, adds under 2 minutes per patient, auditable change log, no new hardware, HIPAA compliant.

- OR supply availability need
  Constraints: Sanitizable surfaces, survives common disinfectants, no RF interference, setup under 5 minutes per room turn, visible from 2 meters, integrates with existing pick-lists.

Step 7: Score all 12 and reduce to 10

Create a small table and score each of the 12 from 1 to 3 on the four axes below. Add the scores to get a total out of 12.

- **Feasibility now**: skills, access, and time you actually have.

- **Strategic fit**: your interests and long-term goals.

- **Effort**: estimated lift to run an initial test.

- **Impact**: plausible magnitude if solved well.

Cut the **bottom 2** totals. Keep **10**. If there is a tie at the bottom, cut the one with weaker problem clarity.

Example row

- Need: A way to ensure accurate medication reconciliation in elderly inpatients to prevent adverse drug events
  Feasibility 3, Strategic fit 3,
  Effort 2, Impact 3, **Total 11**

Quick worksheet templates

Triad to Need (copy for each item)

- Problem:

- Population:

- Outcome:

- Need statement: A way to ... in ... to ...

Readiness scoring (12 rows)

Need | Feasibility (1–3) | Strategic fit (1–3) | Effort (1–3, lower effort = higher score) | Impact (1–3) | Total

Anchor this

You are not choosing a project. You are creating a **clear set of 10** that are framed, scoped, tagged, and pre-thought for constraints. That is exactly what the next challenge expects and will reduce to **3** for focused advancement.

CHAPTER 7: FROM MANY TO ONE: FUNNELING YOUR NEED STATEMENTS

You've done the hard work of gathering friction points, validating observations, and crafting 6 to 10 structured, solution-independent need statements. But if you're like most early-stage builders, you're now feeling the weight of the next step: deciding which one to pursue.

This chapter will walk you through that critical moment. You're not ideating yet. You're funneling: with clarity, not guesswork.

At BMTT, we call this your "pre-commitment phase." It's a period of structured, strategic pruning. The goal isn't to pick your favorite. It's to narrow the field to the 1–3 most aligned, actionable, and high-leverage opportunities based on your time, resources, context, and team.

Why You Still Have Multiple Needs

If you've followed the book so far, you likely have a spreadsheet or notebook filled with filtered need statements: each one potentially worth building around. This is not a problem. In fact, it's a sign of discipline. Most first-time founders prematurely fall in love with one idea and build forward on shaky ground.

Our goal now is to focus your energy where it can create the most value.

STEP 1: DEFINE YOUR FEASIBILITY FILTERS

Before evaluating the need statements themselves, start by defining what makes a project feasible for you at this moment.

Ask yourself and your team:

- What technical capabilities do we already have access to?

- What resources (labs, advisors, funding) can we realistically tap?

- What level of time commitment is available over the next 6–12 months?

- What environments or users can we actually reach for testing and validation?

Create three buckets:

- *High feasibility*: We have both the skillset and access.

- *Medium feasibility*: Doable, but will require new learning or partnerships.

- *Low feasibility*: Exciting, but out of reach given current constraints.

Now, overlay these filters on each need statement.

STEP 2: RANK BY STRATEGIC ALIGNMENT

Feasibility is only one dimension. Next, consider alignment.

Ask:

- Which of these problems energizes me or my team the most?

- Does this need align with the long-term work I want to be known for?

- Is this a problem I'm uniquely positioned to understand or solve?

- Does this connect meaningfully to my lived experience, background, or goals?

Give each need a simple alignment score: High / Medium / Low.

High feasibility + high alignment becomes your "hot zone." These are top contenders.

STEP 3: MAP EFFORT VS. IMPACT

Next, evaluate each remaining need with a simple two-axis model:

- **Effort:** How much time, money, expertise, and coordination would this take to prototype and validate?

- **Impact:** If this were solved well, how many people would benefit and how deeply?

You're aiming for needs that land in the top-right quadrant: *high impact, manageable effort.*

Low effort/low impact is hobby work. High effort/low impact is a trap. High effort/high impact might be better suited for a multi-year research grant or licensing play rather than a student-led startup.

STEP 4: CONSIDER TRACTION PATHWAYS

Now take a moment to think forward. Which needs can gain early traction?

Ask:

- Can I build a quick demo or prototype to test this idea?

- Is there a student pitch competition, grant, or summer accelerator where this fits?

- Are there obvious first users I could reach out to in the next 30 days?

If a need feels powerful but totally untouchable in the short term, shelve it. We're not saying no forever... just not yet.

STEP 5: REDUCE TO 1–3 FINALISTS

At this point, you should have enough context to narrow your list to a maximum of three.

Each of these should:

- Be framed as a clear, solution-independent need statement.

- Align with your time, resources, and skillsets.

- Hold up under scope, feasibility, and impact evaluation.

- Feel exciting to pursue, even when it gets hard.

These are the ideas you'll take into the next chapters. You'll pressure test them against the landscape, map stakeholders, and begin exploring what it would take to win.

CASE REFLECTION: NARROWING THE FIELD

After four weeks of shadowing, late nights, and customer interviews, Nina, José, and Rachel sat at a whiteboard with eight neatly framed need statements and no idea how to pick. Each one had come from real observations, validated through conversations or literature, and scrubbed clean of solution bias. But now the hard part began.

They weren't choosing their favorite. They were choosing their future.

Round 1: Feasibility Gut Check

They rated each need based on what they could realistically take on in the next 6 months.

One idea, reducing implant rejection rates through customized surface coatings, was immediately shelved.

"We don't have access to a tissue lab or histology," Rachel said. "We'd need months of protocol approvals just to start."

Another, focused on reducing diagnostic errors in rare diseases, also landed in the Low Feasibility column. They lacked the connections, access to data, and time required.

By the end of this round, they had narrowed from eight to **five** realistic contenders.

Round 2: Strategic Alignment

Next came the personal filter.

Nina, who had worked in special education before grad school, felt drawn to a project about reducing sensory overload in pediatric MRI scans.

"This is the one I actually want to work on when I wake up."

José leaned toward a mobile triage tool they had brainstormed after seeing bottlenecks in rural clinics.

Rachel had a quiet but growing interest in one about streamlining nurse shift reports. It wasn't flashy, but it resonated.

A few others received lukewarm alignment scores. No one felt strongly against them, but none sparked clear energy either. Rather than eliminate them outright, the team marked those as low alignment and kept them in play if other factors weighed in.

Now down to three top-alignment ideas, plus two maybes.

Round 3: Effort vs. Impact

They drew a quick matrix and placed all five remaining projects on the board.

- The **nursing handoff tool** landed squarely in the high impact, low effort quadrant.

"That's a sleeper," José said. "It's not sexy, but every nurse we talked to brought it up."

- The **pediatric MRI project** was labeled medium effort, high impact. It would take coordination with clinicians, but early ideas were already forming.

- The **triage app** scored high on both effort and impact. Everyone agreed it mattered, but access and testing would be hard.

- The two lower alignment projects landed in either low impact or high effort zones. The group decided to archive those for now.

By the end of this round, three projects remained.

Round 4: Traction Potential

They asked a simple question: What could we actually move on in the next month?

- **Pediatric MRI?** Possible, if they could find the right radiology contact.

- **Nursing handoffs?** Even easier. Rachel's professor ran a simulation lab they could use for testing.

- **Mobile triage?** Tough. It required access to clinical partners they hadn't yet secured.

They put a light circle around the triage tool and labeled it "someday."

Final Picks

Their shortlist:

1. **A way to reduce sensory overload in pediatric MRI procedures to improve scan compliance and patient experience.**

    o   Why? Strong emotional alignment, moderate feasibility, and a clear use case.

    o   Next steps: Interview a radiologist, review existing MRI adaptations, mock up a patient-side interface.

2. **A way to streamline nurse shift handoffs in acute care settings to reduce communication errors.**

- o   *Why?* Underrated impact, very high feasibility, and immediate access to end users.

- o   *Next steps*: Run a workflow study, test low-fidelity scripts in simulation lab, map competing tools.

They paused, then circled both.

"We'll carry both forward for now," Nina said. "But I want to lead the MRI one. I think we have something."

No one objected.

Takeaway

When everything feels like a good idea, structure helps. The goal isn't to eliminate uncertainty; it's to reduce regret. By choosing needs that matched their reality and their drive, the team didn't just narrow the list. They clarified their direction.

FINAL NOTE: LET GO WITHOUT GUILT

You're not abandoning the other ideas. You're creating focus. Many of the greatest founders in history kept notebooks of ideas they didn't pursue: until the timing, team, or technology caught up.

Keep yours too.

One of those ideas may become your next venture, your master's thesis, or your next team's project. But for now, clarity beats optionality.

You're completing the single most overlooked discipline in early innovation: choosing what not to build.

Let's go deeper. Next up: mapping what already exists.

CHALLENGE: From Raw List To Ready Shortlist

You now have 10 real, structured need statements with each one grounded in observed friction and framed without solution bias. But not all of them are right for *you, right now*. This challenge is about using the filters from this chapter to turn that list into a shortlist of 3 project-ready candidates that you can confidently carry forward into landscape analysis and early concept development.

This is not an exercise in perfection. It's an exercise in decision-making under realistic constraints.

Step 1: Create a Project Readiness Matrix

Open a document or spreadsheet and make a simple table with the following columns:

- Need Statement

- Feasibility (High/Med/Low) — Based on your current skills, access, and time

- Strategic Alignment (High/Med/Low) — Based on your interests, values, and long-term goals

- Effort Level (Low/Med/High) — How much work would it take to build and test

- Impact Potential (Low/Med/High) — How much good could this idea do if solved well

- **Traction Potential (Yes/Maybe/No)** — Could you validate this idea in the next 30–60 days?

Fill this out for all 10 of your need statements.

Step 2: Analyze for Your "Hot Zone"

Scan your table and highlight the rows that score:

- **High** in Feasibility
- **High or Medium** in Strategic Alignment
- **Medium or Low** in Effort
- **High** in Impact or Traction

You're not looking for a perfect score across the board. You're looking for *energy zones* (ideas that are exciting, doable, and tied to real users).
Highlight your strongest items and select 3 to carry forward.

Step 3: Articulate Why

For each of your top 3 need statements, write a short (3–5 sentence) explanation:

- Why did you choose this one?
- What makes it a good fit for you or your team?
- What questions do you still need to answer before you commit?

This reflection is important. It will help clarify whether you're chasing novelty, feasibility, or mission, and ensure you're not just picking the one that sounds flashiest.

Step 4: List the First 3 Things You'd Need to Do

For each selected need, write down the **first three things** you'd do to test its viability.

These can include:

- Literature review
- Reaching out to a clinician
- Sketching a use case
- Mapping reimbursement codes
- Searching for competitors
- Interviewing potential users

This step helps make the next chapter tangible; you're starting to map a plan, not just a preference.

Anchor This

Most first-time founders choose an idea based on intuition or excitement. You've now taken a different route: one grounded in feasibility, alignment, and impact.

Your top 3 needs are not just ideas. They're candidates for real-world traction. You'll now move into mapping the landscape by comparing what already exists, what's in development, and where your best idea might actually fit.

Keep this shortlist. It's your compass. And don't throw away the rest; they may come back in ways you can't predict yet.

Let's go make sure your direction isn't just exciting, but informed.

Part III

Landscape Mapping

CHAPTER 8: MAPPING THE LANDSCAPE BEFORE YOU BUILD

A first-principles process for understanding what already exists, why it matters, and where you might fit

You've written a strong, validated need statement. You've resisted the urge to jump into building and have grounded your idea in observable reality. So what now?

At BMTT, the next step is non-negotiable: **landscape characterization**. Before we begin brainstorming, prototyping, or sketching a single concept, we require our teams to conduct a deep, structured analysis of what currently exists. That includes understanding the standard of care, mapping emerging technologies, analyzing what's in the development pipeline, and identifying regulatory hurdles and market dynamics.

This chapter outlines that exact process.

Why is this step so critical? Because the most common mistake early-stage innovators make, especially students and researchers, is assuming that the absence of a solution in *their* environment means it doesn't exist at all. We've seen time and time again that this assumption leads to wasted effort, redundant ideas, and avoidable dead ends.

A physician may say, "We really need a better tool for X," and students jump into solution mode without realizing that tool already exists, has FDA clearance, and has been used successfully in dozens of systems, just not in that physician's hospital. The product may have failed to be adopted for a thousand reasons, none of which have to do with its clinical performance.

The lesson: you must understand the terrain before you build anything new.

STEP 1: UNBIASED DESCRIPTION OF THE STANDARD OF CARE

The first step in your landscape mapping is to describe, as objectively as possible, the current standard of care.

Ask yourself:

- What is the dominant approach currently used to address the problem in your need statement?

- How does it work? Who uses it? Where is it used?

- What outcome is it designed to achieve?

- Who decides to adopt or purchase the solution? Is that the same person who actually uses it day-to-day?

- Why might users stick with the status quo, even if it seems inefficient? Does the current approach save time, avoid complexity, or reduce training burden in ways that make it "good enough" for them?

Your job here is not to critique or solve. It is to observe and document. Describe what is done, by whom, and to what effect. Be careful not to embed your opinions yet. This step is about learning, not judging.

Why it matters: A clear understanding of the standard of care helps you see whether your idea is truly novel, or whether it is simply a less efficient version of what is already widely available. It also highlights hidden barriers to adoption: sometimes the reason a tool is not used has less to do with performance and more to do with time-cost, workflow friction, or the fact that buyers and users are not the same people. By noticing these dynamics early, you build a foundation for more realistic innovation work in later steps.

STEP 2: SCAN FOR FRESH TECHNOLOGY

Once you've mapped the dominant players, shift your focus to **emerging technologies.** These are the newer devices, processes, or digital tools that have been launched in the past 3–7 years and are intended to improve on the standard of care.

Your goal is to identify:

- How these innovations work

- What specific advantage they claim (better accuracy, faster workflow, fewer complications, etc.)

- What markets they're targeting

- What limitations they still face (adoption, reimbursement, usability, etc.)

Avoid hype and marketing language. Focus on function and fit.

Sources to explore:

- LinkedIn and company websites

- PubMed or Google Scholar for new clinical validations

- Patent databases (Google Patents)

- CrunchBase

- News articles and MedTech publications

- FDA's 510(k) or De Novo databases

Why it matters: This stage helps you benchmark what's truly innovative versus what's already in motion. You're identifying where your idea overlaps with, improves upon, or diverges from what's already on the market. You may discover you're not as early as you thought, or that your opportunity lies in making something more accessible, affordable, or usable, rather than more novel.

STEP 3: ANALYZE THE GAPS IN BOTH

Now comes the **critical analysis** phase. This is where you compare the standard of care and fresh technology, not to declare a winner, but to **map the gaps** in both.

Ask:

- What are the persistent shortcomings of the standard of care?
- What do newer technologies claim to solve, and where do they fall short?
- Where is adoption failing despite technical success?
- What are the known user pain points, maintenance issues, or learning curve barriers?

This is where many of the most compelling opportunities reveal themselves: not in what's *broken*, but in what's *underutilized*, *mismatched*, or *incomplete*.

Why it matters: If your idea doesn't outperform or fill a clear gap between existing options, it will struggle to justify itself in an investor pitch, a purchasing decision, or a hospital workflow. Identifying these gaps early also helps you refine your value proposition before you waste time building.

STEP 4: RESEARCH THE PIPELINE, BUT KEEP IT INTERNAL

This step is about identifying technologies **still in development**, whether in stealth mode, undergoing trials, or early in the regulatory pipeline.

- **Use these tools:**

- Google Patents: Search by keywords, companies, or inventors

- ClinicalTrials.gov: Look for trials involving your problem space

- SBIR/STTR grant databases: Explore funded research and device development

- Press releases and startup competitions

Important: While this research is essential for *your* internal strategy, it should not dominate your investor pitch. These technologies are unproven, and your focus should remain on validated pain points and currently available solutions.

Why it matters: Knowing what's coming helps you anticipate your competition. It informs how quickly you need to move, how different your idea really is, and whether you should pivot now instead of later. You're not looking to out-compete the pipeline; you're trying to make sure you don't walk into it blindly.

STEP 5: IDENTIFY REGULATORY HURDLES AND CASE STUDY CHALLENGES

Regulatory clearance is often the single largest barrier to entry in healthcare. A brilliant idea that doesn't fit within a regulatory pathway (or that takes seven years and $10 million to validate) may not be a startup at all. It might be a research project.

Your task: Look into how competitors navigated regulatory approval.

Ask:

- Did they pursue a 510(k), De Novo, or PMA? *(See quick reference below.)*

- Did they run clinical trials? If so, what were the phases and endpoints?

- Were there delays, rejections, or resubmissions?

- Did reimbursement affect their rollout?

Quick Reference: Regulatory Pathways

- **510(k):** The most common pathway for devices that are "substantially equivalent" to an existing product. Usually faster and less costly.

- **De Novo:** For novel, low-to-moderate-risk devices without a clear predicate. Creates a new regulatory category.

- **PMA (Premarket Approval):** The most rigorous pathway, required for high-risk Class III devices. Demands extensive clinical data and long timelines.

Where to look:

- FDA 510(k) database and approval letters

- Medtech Insight or similar trade publications

- SEC filings from public companies

- Case studies in accelerator portfolios or academic presentations

Why it matters: This isn't just about avoiding red tape. Understanding regulatory precedent helps you map development timelines, plan budgets, and communicate more clearly with funders. It also helps you understand if a "simple" idea is actually entangled in decades of compliance complexity.

STEP 6: BUILD YOUR MARKET MAP

The final step in this analysis is to **visualize** where your idea sits relative to the competitive landscape. This doesn't need to be polished or pretty yet, but it needs to be real.

What to include:

- The current standard of care (center of the map)

- Emerging or recently launched technologies (positioned around it)

- Pipeline technologies (at the perimeter, with timelines if available)

- Regulatory milestones (FDA status, CE markings, etc.)

- Your perceived opportunity (drawn with full context, not as a blank space)

Why it matters: A market map helps you avoid redundant thinking. It shows you where there is genuine whitespace and where the "gap" you were chasing is actually already being filled.

CASE REFLECTION: THE MAP BEFORE THE MOUNTAIN

Nikhil's team was sure they had something. Their need statement, *'a way to reduce burnout among surgical residents during night shifts to improve mental health and patient safety'*, had been validated across interviews at three institutions. It checked all their boxes: relevant, timely, and personally resonant.

They almost skipped the landscape phase.

"There's no product for this," one teammate said. "Everyone just complains about it."
But their program required a full landscape analysis before moving forward.

Standard of Care
Their first task was describing how the problem was currently handled. After a few calls and some digging, the team realized there were protocols already in place: call schedules, wellness check-ins, and institutional mental health resources.

"It's not that nothing exists," Nikhil noted. "It's that people don't use what's already there."

The standard of care was fragmented, inconsistent, and rarely optimized.

Current Technologies
Then came the competitive scan. A few hours into their research, the team found at least four tools on the market targeting clinician burnout: two apps for wellness tracking, a platform offering peer check-ins, and a scheduling tool that used predictive analytics to reduce fatigue.

Most weren't focused on residents specifically. Some had limited adoption or low user engagement. But they existed.

"This just went from 'nothing is out there' to 'why isn't what's out there working?'" Nikhil said.

Gap Analysis
With their assumptions challenged, they shifted gears. The gap wasn't a lack of technology. It was that the existing solutions weren't designed around the workflow realities of surgical residents. They weren't tailored to time constraints, cultural norms, or trust barriers. That reframed everything.

Pipeline Awareness
ClinicalTrials.gov showed a few pilot programs testing behavioral nudges and shift adjustments. A recent SBIR grant funded a team building a VR-based stress training module for emergency medicine. The team took notes but didn't feel crowded out yet. They saw an opening: not to invent a new tool, but to integrate and adapt.

Regulatory Precedents
They scanned FDA records and found that most similar tools were categorized as wellness apps or low-risk SaMDs. No clinical trials had been required. That reduced their concern about long timelines or heavy compliance work.

Market Mapping
On a rough slide, they placed the fragmented standard of care in the center, surrounded by commercial tools, pilot studies, and university programs. Their proposed idea sat between categories: part scheduling aid, part behavioral support. Not a replacement. A reinforcement.

"We're not competing with any one thing," Nikhil said. "We're stitching together a system that already exists but doesn't work in practice."

Takeaway
The landscape analysis didn't kill their idea. It sharpened it. By understanding what had failed, what had been ignored, and what had already tried to solve the problem, the team saw where they could actually add value. Their idea was no longer built on frustration alone. It was built on context.

FINAL THOUGHTS: YOU CANNOT SKIP THIS

At BMTT, this process is the foundation of everything that comes next. It's how we decide whether a project deserves resourcing, whether a need statement should be refined, or whether we're about to build something that no one needs or wants.

It is tempting to think you're different. That your idea is so intuitive and novel that it couldn't possibly exist already. But the odds are against that. And your best defense is research.

Map the landscape first. Not to kill your idea, but to anchor it in reality.

Once you've done this, you're ready to dive into stakeholder mapping, regulatory positioning, and workflow integration.

CHALLENGE: The Landscape Gauntlet

You've narrowed your funnel. You now have 3 promising need statements, each validated and vetted. But you haven't chosen yet and you shouldn't. Not until you've mapped the full terrain.

This challenge will guide you through a structured landscape comparison, revealing where the real opportunity lies and helping you avoid costly redundancy or blind spots before they happen.

By the end of this, you'll have made your first major commitment: picking a problem to build around.

Step 1: Landscape Reports for Each Candidate

For each of your 3 remaining need statements, complete the following:

A. Standard of Care Brief

- What is currently used to address this need?

- How effective is it? Who uses it? Where?

- Is it procedural, device-based, behavioral, or administrative?

Format: One short paragraph or a bullet list.

B. Competitor Summary

List at least 2 current technologies or services addressing this need.

- Name

- How they work

- Value proposition

- Limitations or failure points

Format: Table or bullet format for quick comparisons.

C. Pipeline Watch

List 1–2 pipeline technologies, including:

- Company or research group

- Status (trial phase, stealth mode, grant-funded, etc.)

- Claimed benefit

- Estimated time to market (if available)

Note: Use Google Patents, ClinicalTrials.gov, SBIR databases, or news outlets.

D. Gap Synopsis

Write 3–5 bullets describing what the unmet need still is (even with existing solutions).

- Is it usability?

- Price?

- Adoption?

- Workflow fit?

- Technical performance?

This is your unique wedge: the thing that remains unsolved.

Step 2: Regulatory Pathway Snapshot

For each candidate:

- Identify its likely product classification (e.g., Class II device, SaMD, diagnostic, consumer tool).

- Search for 1–2 relevant regulatory precedents.

- Estimate clearance or approval pathway (e.g., 510(k) with predicate X, De Novo, PMA).

Optional but encouraged: Check reimbursement context (CPT codes, DRGs, or known payer resistance).

Step 3: Market Map Sketch (Rough)

You don't need graphics software. Just draw it on paper or in a slide. Include:

- The current standard of care (center)

- Current competitors (grouped nearby)

- Pipeline (noted on the perimeter)

- Your candidate's position (showing where it's similar and where it diverges)

- Optional: Include visual tags for regulatory burdens or adoption barriers

Step 4: Comparative Matrix

Create a comparison chart across all 3 candidates using these categories:

Candidate	Maket Gap Clarity	Competitive Crowding	Regulatory Burden	Feasibility (Team Fit)	Impact Potential	Adoption Likelihood	Your Excitement Level
[Name]	High/ Medium/ Low	High/ Medium/ Low	High/ Medium/ Low	High/ Medium/ Low	High/ Medium/ Low	High/ Medium/ Low	High/ Medium/ Low

Score each honestly. You're not proving anything yet; you're getting clarity.

Step 5: Make the Call

Review your landscape reports and matrix. Then write a 2-3 paragraph justification for which need statement you are officially choosing to move forward with.

Answer:

- Why this one over the others?//
- What do you still need to learn before building?
- What would success look like if this worked?

This is your commitment checkpoint. From here on, your work will orbit this decision.

Anchor This

Most ideas fall apart not because they're bad, but because they were selected too early, without the information to make that choice wisely.

You now have that information.

You've mapped what exists, what's emerging, what's failed, and what still needs solving. You've compared feasibility, excitement, adoption, and opportunity. You're not chasing what's shiny. You're choosing what's real.

Save all of your work; it will become the raw material for your pitch deck, partner conversations, and product design.

Now: commit, then let's start building.

CHAPTER 9: UNDERSTANDING THE TERRAIN

Stakeholder mapping, regulatory snapshots, and workflow integration

Once you've mapped the technological and regulatory landscape surrounding your clinical need, you're one step closer to building something real. But before you start developing a solution, or even selecting one, you must understand **who** will interact with it, **how** it will function in practice, and **what barriers** will prevent it from being adopted.

We call this understanding the terrain.

In many ways, this step is where the idea stops being theoretical. A product that looks great in a vacuum can quickly fall apart when introduced into a real clinical workflow. Likewise, a tool that works in one setting may be completely unusable in another because of staffing norms, billing constraints, or cultural resistance.

This chapter will walk you through three interconnected exercises that must happen before you build:

1. Stakeholder mapping

2. Regulatory pathway assessment

3. Workflow and system integration planning

Together, these form the foundation for a successful go-to-market strategy, but more importantly, they inform the actual design of your solution.

1. STAKEHOLDER MAPPING: WHO TOUCHES THE PROBLEM?

In healthcare, no product exists in a silo. Even the simplest device or software tool touches multiple stakeholders: patients, clinicians, nurses, administrators, techs, payers, and often procurement staff. If your solution only works for one of them, but creates friction for the others, it will struggle to get adopted.

Start by Listing the Stakeholders

Your first task is to **map every role that interacts with the need** you've identified. Start with those closest to the problem, and then expand outward.

For example, if your need statement is:

'A way to reduce catheter-associated infections in post-op patients to improve recovery outcomes'

Your stakeholder map might include:

- Bedside nurses (monitoring and flushing the catheter)
- Physicians (ordering and overseeing treatment)
- Patients (experiencing the complication)
- Infection control officers (tracking infection rates)
- Procurement staff (deciding what brands or products to purchase)
- Clinical educators (training staff on correct use)
- Hospital administrators (concerned about penalties tied to infection rates)

- Insurance payers (deciding on reimbursement tied to complications)

This list should be specific and complete, not just "doctors" and "nurses," but which types, in which environments, with what responsibilities.

※ BMTT Tip + Case Study: Talk Early, but Talk to Everyone

This is a great time to begin talking with at least one person in each stakeholder group. Early conversations can help you check assumptions, better understand motivations, and begin mapping out the ecosystem. But don't mistake a few enthusiastic conversations for full validation. Before you spend serious time or capital building an MVP, it is essential that you've spoken with multiple representatives from every stakeholder group involved in the care pathway, purchase process, and usage of your product.

We cover this in depth in the upcoming *Customer Discovery* chapter, where we also share multiple case studies of what happens when you don't.

A striking example: Pfizer's $2.8 billion failure with Exubera, a needle-free insulin inhaler. The product was built on a perfectly reasonable assumption that patients dislike needles and would embrace a more convenient alternative. What Pfizer failed to do was test that assumption with real users. Had they conducted even a modest customer discovery effort with insulin-dependent diabetics, they would have learned that few were willing to carry around a bulky, bong-like inhalation device. Exubera flopped within a year of launch, and it remains one of the most expensive product failures in pharmaceutical history.

At BMTT, this problem is even more acute because many of our clinical needs come directly from the source: healthcare professionals. While well-intentioned, these practitioners are sometimes unaware of current technologies, are not always following standardized protocols, or were undertrained on device usage. This creates false perceptions of need, which student teams, eager to help, may act on prematurely.

So, yes, talk to stakeholders early. But validate widely. Your solution must fit not just the *idea* of the problem, but the system in which that problem lives.

Categorize Stakeholders by Role

Next, map each stakeholder into one of four buckets:

- **End users**: Who physically interacts with the product?

- **Decision-makers**: Who authorizes the use or purchase of the product?

- **Influencers**: Who shapes opinions and protocols (e.g., respected clinicians or administrators)?

- **Beneficiaries**: Who benefits most directly from the improved outcome?

Keep in mind that **your buyer is rarely your user**, especially in healthcare. In many cases, the person you're solving for has no purchasing authority. And the person who buys your product may care more about risk, reimbursement, or reputation than functionality.

This disconnect is one of the main reasons otherwise brilliant solutions fail to gain traction.

2. REGULATORY SNAPSHOTS: WHAT HURDLES LIE AHEAD?

Even the most elegantly designed product can be stopped cold by regulatory constraints. That's why one of your earliest responsibilities is to determine **what kind of product you're building from a regulatory perspective**, and what pathway it will need to follow.

Step One: Determine Your Product Category

Is your eventual solution likely to be a:

- Class I, II, or III medical device?

- Drug-delivery system?

- Digital health software?

- Diagnostic tool?

- Combination product?

- Consumer wellness tool?

- Decision support system or AI model?

This classification determines your likely **regulatory pathway**, **evidence burden**, and **timeline to approval**.

Step Two: Identify Precedents

Use the FDA's 510(k) database, De Novo approvals, or PMA databases to search for similar products.

Ask:

- How were they classified?

- What kind of testing did they require?

- Did they have predicate devices, or were they entirely novel?

- How long did it take to reach clearance or approval?

If you're building a digital tool, check whether it qualifies as **Software as a Medical Device (SaMD)**. If it makes clinical claims or influences medical decision-making, it may be regulated, even if it's "just an app."

Step Three: Identify Global Considerations

If you plan to sell outside the U.S., learn about CE marking, MDR in the EU, or equivalent pathways in Canada, Japan, and Australia. Regulations vary significantly across borders, and your product's classification may shift accordingly.

BMTT Tip: You do not need to become a regulatory expert overnight. But you *do* need to understand the pathway well enough to plan around it. If your product will require a PMA, it could take 3–7 years and millions in clinical trials. If it's eligible for a 510(k) clearance with a known predicate, the timeline and cost could be far more manageable.

---

3. WORKFLOW INTEGRATION: WHERE DOES IT FIT?

The last and often most neglected part of terrain mapping is understanding **how your product will integrate into real-world**

clinical workflows. Too many student teams design for the problem, not the environment.

The Questions to Ask:

- At what point in the care pathway will your product be used?

- Who initiates its use? Who maintains it? Who documents it?

- What current steps will it replace, add to, or eliminate?

- How will it interact with other tools, systems, or protocols?

- What risks does it introduce technically, legally, or operationally?

This step is crucial because even small interruptions to workflow can prevent adoption. Clinicians are time-starved and trained in routines. If your tool takes 45 seconds longer to use or requires re-training, you'll need to prove that the outcome improvement is worth that tradeoff.

Consider: Hospital and Outpatient Differences

Workflow integration looks vastly different across settings:

- In a **hospital**, your product may need to interface with EMRs, pass IT and infection control checks, and be maintained by nursing staff.

- In an **outpatient clinic**, your product may be handled directly by a provider with no supporting team.

- In **home care**, patient ease-of-use and clarity of instruction are paramount.

※ **BMTT Tip: Run a "Day in the Life" Simulation**
We often ask teams to simulate a full day of use for their product: who interacts with it, when, for how long, and in what setting. This helps uncover friction points early and ensures your development process is aligned with the people you're actually serving.

CASE REFLECTION: REALITY CHECK IN THE ICU

Aisha's team had momentum. Their need statement, '*a way to reduce delays in ICU patient transfers to step-down units to improve bed turnover*', had been through customer interviews, had data to back it, and even had a few "I'd use that tomorrow" quotes from clinicians.

They thought they were ready to brainstorm solutions. But their advisor hit pause.

"Before you sketch a single wireframe," she said, "map out the terrain. If it can't survive the system, it won't survive the market."

Stakeholder Mapping

They expected it to be simple: nurses request a transfer, bed availability is confirmed, and patients are moved.

But once they started mapping stakeholders, it unraveled.

- **Charge nurses** coordinated transfers but were juggling staffing ratios.

- **Case managers** were responsible for insurance authorization.

- **Transport teams** were often backed up for hours.

- **Receiving unit nurses** sometimes declined patients based on acuity.

- **ICU physicians** had the final say, but were often unavailable during shift changes.

What seemed like a single-step delay was actually a six-person relay with no clear baton handoff.

"We thought the problem was communication," Aisha said. "It's actually ownership."

Regulatory Snapshot

They figured a simple coordination app could solve it until they learned anything interfacing with patient records would likely qualify as SaMD.

A quick scan of FDA filings showed that even communication tools, if used to inform clinical decisions, often required clearance. They found a similar handoff tool that had been stuck in the De Novo process for over a year.

"If we can't prove clinical utility, we may be seen as an IT burden rather than a medical device," one teammate realized.

They shifted their thinking: maybe the tool doesn't suggest decisions, just tracks status.

Workflow Simulation

They mapped a typical day in the ICU and quickly hit a snag: no one had 30 seconds to spare.

During rounds, clinicians were in motion. Nurses were managing meds, techs were on phones, and administrators were chasing forms. If their solution required active updates, it wouldn't get used.

They revised the idea: instead of adding steps, what if they just made the current status visible to everyone?

"Think Kanban board, not dashboard," one teammate suggested: in other words, a simple shared board where status is visible at a glance, like sticky notes on a wall, not another data-heavy screen.

A Day in the Life

They wrote a narrative for Lauren, a night-shift ICU nurse.

Lauren had eight patients, two at risk of crashing, and one scheduled for transfer. The transfer was delayed three hours, not because a bed wasn't ready, but because the step-down unit hadn't been notified in time, and once they were, no one followed up.

Lauren didn't forget. She just prioritized survival.

For their solution to work, it had to think like Lauren did. It had to *remember* for her.

Takeaway

The terrain mapping didn't just refine their direction; it rewrote it.

Their initial concept, a decision-support transfer app, was dropped. What emerged was a passive alert system tied to existing scheduling tools, designed to nudge the right person at the right time.

It wasn't glamorous. But it fit.

PUTTING IT ALL TOGETHER

These three exercises (stakeholder mapping, regulatory snapshots, and workflow integration) will give you a full picture of the terrain you're stepping into. They are not just boxes to check. They are part of the **design process**.

A well-framed solution is one that:

- Addresses a validated need
- Fits into a real workflow
- Navigates a known regulatory path
- Satisfies the person who uses it *and* the one who pays for it
- Avoids unnecessary friction or scope creep

If you skip this terrain-mapping process, you risk building something that works in theory but fails in the world.

If you take the time to do it right, you will build something that lives.

In the next chapter, we'll tackle another foundational piece of your strategy: how to define your market clearly, convincingly, and without resorting to made-up numbers.

Let's talk about **who pays, who benefits, and how to calculate TAM, SAM, and SOM without needing a finance degree**.

CHALLENGE: Your Problem, In The Wild

Congratulations. If you've made it this far, you've done something most founders never do: you've resisted the temptation to build too early. You've clarified a real, observable, high-impact need and chosen one to pursue with purpose.

Now comes the work of embedding that problem in its actual environment.

This challenge will help you map the system around your need: the people it touches, the rules that govern it, and the workflows that could make or break your solution.

You are no longer exploring what to build. You are learning where it must live.

Step 1: Stakeholder Web

Create a complete stakeholder map for your chosen need. Include:

- **Primary Users** (the person or people who will actually interact with your future solution)

- **Decision-Makers** (the person or group who authorizes or purchases the product)

- **Influencers** (anyone who shapes decisions (clinical leads, respected peers, administrator))

- **Beneficiaries** (those who are helped by the product but may never touch it)

For each stakeholder:

- List their role

- What they currently do related to the problem

- What matters most to them (speed, safety, reimbursement, simplicity, etc.)

- Any known friction they already face

This web will guide your customer discovery, design, and communication strategy.

Step 2: Regulatory Snapshot

You're not filing anything yet, but you need to know the terrain.

For your need, identify the most likely product category (we know you do not have a solution yet; that is by design!):

- Device (Class I, II, or III)?

- Diagnostic?

- Drug-device combo?

- Digital tool with clinical claims (SaMD)?

- Consumer wellness?

- Other?

Then find at least **one regulatory precedent**:

- A similar device or tool that has been cleared or approved

- What pathway it followed (e.g., 510(k), PMA, De Novo)

- How long it took

- What type of testing or validation it required

Write a short paragraph summarizing what you learned and what it implies for your timeline and risk.

Step 3: Workflow Fit Simulation

This is where paper plans meet clinical reality.

Using your stakeholder map and observations so far, walk through the exact sequence of events where your solution would be introduced.

Answer:

- When in the care journey does this need arise?

- Where would your solution be used?

- Who would initiate its use? Who would maintain or document it?

- What steps would it replace or add?

- What other systems would it interact with (software, devices, people)?

Then answer the harder question: **where could it fail to fit?**

List at least 3 risks to integration:

- Workflow friction

- Training needs

- Regulatory hesitancy

- User resistance

- Cultural misalignment

Write these down now so you can design around them later.

Step 4: Day in the Life

Choose a specific stakeholder (preferably the **primary user** or **primary decision-maker**).

Now, write a one-paragraph narrative of a normal day in their life, including:

- Where and when your problem shows up

- What they're juggling at that moment

- What trade-offs they're managing

- What makes them ignore the problem (or work around it)

- What they'd need to see, feel, or experience to adopt a new solution

This is how you stop building in the abstract and start building with context.

Anchor This

You now know something many founders avoid until it's too late: your idea must survive contact with a system.

You've mapped the people, the red tape, the routines, and the resistance. That knowledge will save you months and help you build a solution that actually lands.

Keep these notes close. Every design decision from here forward should echo what you've just learned. And when the time comes to test, pitch, or build, this terrain map will remind you what matters most: not elegance, not features, but fit.

CHAPTER 10: MARKETS: WHO PAYS, AND WHO BENEFITS?

TAM, SAM, and SOM made simple, whether you're in a business school or a biology lab

By now, you've defined a real clinical need, mapped the landscape of existing solutions, and explored how your future product might interact with stakeholders and workflows. But before you get too far down the design or development path, you need to stop and ask one deceptively simple question:

Is there a market for this, and more specifically...

- Who benefits if this problem is solved?

- Who would pay for the solution?

- And how many of those people or institutions actually exist?

At BMTT, we often remind our student teams that success in innovation isn't just about solving a problem. It's about solving a problem *for someone who is both willing and able to pay for the solution.*

This chapter will walk you through how to think about your market clearly, systematically, and without relying on inflated projections or hand-waving business jargon. Whether you're a premed student, a biomedical engineer, or a first-time founder, understanding TAM, SAM, and SOM is your key to building something that not only works, but survives.

Why Market Sizing Matters

Your market is not "everyone who could benefit from your product." That's a fantasy. Your real market is shaped by clinical practice, reimbursement models, regulatory access, geographic constraints, and user behavior.

Market sizing helps you:

- Know whether your idea is worth pursuing
- Forecast the resources you'll need to succeed
- Communicate with investors or grant reviewers
- Set realistic short- and long-term goals
- Avoid building for a market that doesn't exist

Let's walk through the three tiers of market analysis: **TAM**, **SAM**, and **SOM**.

1. TAM: TOTAL ADDRESSABLE MARKET

The TAM, or **Total Addressable Market**, is the absolute *largest possible* market your product could ever serve, assuming no barriers to adoption, reimbursement, or regulation.

Think of TAM as the ceiling. It represents the maximum revenue opportunity if your solution was adopted globally by everyone who needed it.

⊗ **Example:**

Victor is looking to develop a product to improve insulin adherence in Type 1 diabetics. His first step is to find out how many people in the world live with Type 1 diabetes. According to the World Health Organization (WHO) 2025 report, there are about **9 million people worldwide diagnosed with Type 1 diabetes**.

Victor then multiplies this by the projected annual cost of his solution. If his product would cost **$1,000 per patient per year**, his TAM is:

TAM = 9 million patients × $1,000 = $9 billion per year

This gives Victor the ceiling: the absolute maximum his market could ever be worth if every patient in the world used his product.

How to Estimate TAM:

- Use epidemiological data from sources like the CDC, WHO, or national databases

- Multiply the total number of relevant patients/procedures/devices by your anticipated price point or cost savings

- Avoid fluff: do not include loosely related conditions or marginal use cases just to inflate your numbers

※ **BMTT Tip:** If you're targeting a system-wide workflow or device improvement, your TAM might be based on **procedures** rather than patients. For instance, a new surgical tool might have a TAM based on the number of procedures per year where that tool would be used.

2. SAM: SERVICEABLE AVAILABLE MARKET

Your SAM, or **Serviceable Available Market**, is the portion of the TAM that your solution could realistically serve based on current access, regulation, and clinical application.

SAM answers the question:
"Who can I serve, given my current scope?"

✗ **Using the same insulin example:**
If your inhalable insulin device is only cleared for use in the United States, and only for patients aged 18–65 with no pulmonary complications, your SAM is:

- U.S.-based Type 1 diabetics aged 18–65 without contraindications

SAM = Subset of the total population × Expected product use × Market pricing

You might also adjust SAM based on clinical setting. If your tool is only viable in outpatient clinics, not hospitals, that constraint must shape your numbers.

How to Estimate SAM:

- Use filtered population data (e.g., Medicare/Medicaid databases, EMR usage data, procedure codes)

- Apply known constraints like age, geography, diagnosis severity, or clinician type

- Factor in regulatory approvals, reimbursement access, and clinical protocols

---

※ **BMTT Tip:** SAM is the most useful number for mid-stage planning. It helps you set achievable near-term goals and understand how your target market intersects with reality.

---

3. SOM: SERVICEABLE OBTAINABLE MARKET

SOM, or **Serviceable Obtainable Market**, is the **realistic** portion of the SAM that you can capture in the early years of your venture.

SOM answers the question:
"Who will we actually reach, given our current resources, team, partnerships, and go-to-market plan?"

This is where real strategy comes into play. SOM is not a spreadsheet exercise. It's a credibility exercise. Investors, grant reviewers, and partners want to see that you understand the difference between ambition and execution.

✖ **Continuing the example:**

If your product is initially launching in five states, with two pilot sites, and you only have bandwidth to market to large endocrinology practices, your SOM might look like:

- 50,000 potential patients in early target regions
  × 30% coverage of target clinics
  × %15 conversion rate in first year
  × $500/year per patient = ~$1.1M in Year 1 revenue potential

How to Estimate SOM:

- Identify your initial geographic footprint

- Project realistic outreach, adoption, and conversion **rates**

- Use real benchmarks: What have comparable products achieved in year 1–2?

- Refine this number often; it's your operational north star

※ **BMTT Tip:** If you're pursuing non-dilutive grants like SBIR or STTR, funders will scrutinize your SOM even more than your TAM. Show them you understand *who* you're targeting, *why* they'll say yes, and *how* you plan to reach them.

USING TOP-DOWN AND BOTTOM-UP MARKET SIZING: A REALITY CHECK

When sizing your market, it's easy to fall into the trap of only using a **top-down** approach. This is the most common method for first-time founders because it's fast and accessible: you look up the total market size from a report or database and estimate what fraction you might capture.

⊠ Example (Top-Down):
"The global wound care market is worth $25 billion. If we capture just 1%, that's a $250 million opportunity."
That number may look exciting, but it's not credible on its own.

To strengthen your case, you should also run a **bottom-up analysis**. This approach starts from the ground: looking at real-world usage patterns, procedure volumes, sales channels, and expected pricing.

⊠ Example (Bottom-Up):
"There are 200,000 relevant surgical procedures annually in our launch region. We estimate reaching 10% of those within three years, with a product price of $400 per case, leading to a $8 million obtainable market."

Why You Need Both

Running both top-down and bottom-up models is more than a credibility play; it's a sanity check. If your two estimates are wildly misaligned, that tells you something important:

- Are you overestimating what you can realistically capture?

- Are you underestimating the number of addressable users?

- Are your assumptions about adoption rate, pricing, or access wrong?

Where the numbers align, you gain confidence. Where they diverge, you gain insight.

※ **BMTT Tip:** Always test your bottom-up assumptions by talking with people who actually manage clinical volume or procurement. Ask them how many procedures are done each month, what percentage of providers might be early adopters, and what price points have caused friction in the past.

WHO PAYS? WHO BENEFITS?

Market sizing is only half the story. The other half is **understanding your economic stakeholders**.

In healthcare, the person who benefits from a product is often not the person who pays for it.

You need to map:

- **End users** (e.g., nurses, patients, caregivers)

- **Purchasers** (e.g., hospital procurement, private clinics, insurers)

- **Payers** (e.g., CMS, private insurers, out-of-pocket)

- **Gatekeepers** (e.g., IT departments, medical boards, ethics committees)

If your product requires **prior authorization**, competes for formulary placement, or adds steps to a workflow, even a clear clinical benefit may not be enough.

Key Questions:

- Who has the **budget** for this product?

- Who has the **authority** to say yes?

- Who will push for adoption, and who might resist?

- Will this be billed under existing codes or require new reimbursement pathways?

※ **BMTT Tip:** You don't need all the answers on Day 1. But you do need to start asking the right questions. Early interviews should include at least one **procurement officer**, **billing expert**, or **practice manager**... someone who understands the real-world economics of technology adoption.

HOW TO VISUALIZE THIS IN A PITCH OR REPORT

When presenting your market analysis to others (investors, mentors, reviewers), keep it simple and honest.

A great market slide:

- Shows TAM, SAM, and SOM clearly

- Breaks down the assumptions behind each number

- Differentiates between potential and reality

- Includes a "Who Pays vs. Who Benefits" table or diagram

- Leaves space for skepticism and follow-up questions

A bad market slide:

- Uses massive round numbers with no cited sources

- Claims 1% of a billion-dollar market as a "conservative estimate"

- Lists the global diabetic population without filters or context

- Assumes overnight adoption by payers or providers

CASE REFLECTION: THE $0 SOLUTION

Jamal's team was riding high. Their idea, a low-cost, modular tool to help community hospitals reduce unnecessary overnight admissions, had real clinical traction. Multiple physicians had called it "game-changing." Even the pilot site's chief medical officer gave them a thumbs-up.

It was one of the cleanest need statements they'd ever worked with.

But two weeks before their end-of-semester pitch, something didn't add up.

"Why are none of your stakeholders listed under purchasers?" a visiting VC asked during a dry run.
"Who pays for this?"

They blinked.

"We assumed the hospital would."

TAM? Great. SAM? Solid. SOM? …Missing

Their TAM looked promising: millions of unnecessary admissions each year in U.S. hospitals, costing billions.

Their SAM: **filtered to mid-sized hospitals in rural and underserved areas with similar bed utilization patterns. Smart.** But when it came to SOM, their actual Year 1 target, they had no beachhead, no path to pilot funding, and no buyer with budget authority.

The Assumption Gap

It turned out their primary users, emergency physicians, loved the idea. But they had zero purchasing power.

Hospital CFOs, the real decision-makers, weren't nearly as excited.

One even told them:

"Fewer overnight admissions might improve care, but it doesn't help our financials unless CMS is incentivizing it."

In fact, reducing admissions would LOSE the hospital money under the current reimbursement model.

Jamal's team had built something that saved the system money but not the hospital. Their SOM was effectively $0.

The Pivot

They reworked the pitch: not the product, but the **value framing**.

They mapped who BENEFITED (patients, CMS, clinicians) and who PAID (currently no one). Then they hunted for aligned incentives.

- Could they tie the tool to **bundled payment** programs?
- Could they sell it to **accountable care organizations (ACOs)**?
- Could CMS or state Medicaid pilots fund adoption?

Eventually, they found a viable wedge: hospitals in value-based care arrangements WERE interested because their incentives were aligned.

Their new SOM wasn't huge, but it was real.

"We went from building for the system to building for a payer-aligned buyer," Jamal said.
"Same product. Different market. And now we actually have a shot."

FINAL NOTE: MARKET STRATEGY IS A LIVING DOCUMENT

Your understanding of the market will evolve constantly. New competitors will emerge. Reimbursement codes will change. Clinical practice patterns will shift.

What matters is that you remain adaptive and evidence-driven. Market analysis is not a one-time pitch deck exercise. It's a continuous process that guides your development, your partnerships, and your fundraising strategy.

Market sizing gives you your target. Customer discovery tells you whether that target is real.

CHALLENGE: Building a Market That's Real, Not Just Big

You now have a single, high-impact, validated need that sits at the center of everything you've built so far. That's no small feat. Most aspiring founders never make it this far without spinning into half-baked ideas or inflated narratives.

Now the real test begins: **does this problem live inside a real, reachable, and valuable market?**

This challenge will help you answer that question with numbers, not guesses. By the end of it, you'll have the bones of a pitch-ready market strategy that actually holds up under scrutiny.

Step 1: TAM — Define the Ceiling

Your Total Addressable Market is the largest possible version of your opportunity.

Use epidemiological, procedural, or systems-level data to estimate:

- How many people experience the problem?
- How often? (One-time, recurring, chronic, procedural, etc.)
- What would your price or savings per user be?

Then calculate your TAM:

- *# of relevant people or procedures × Estimated annual price or savings*

Write a 3–5 sentence justification citing your data source(s). No fluff.

Step 2: SAM — Define the Realistic Reach

Now ground your numbers.

- Given your geography, regulatory pathway, access constraints, and clinical scope:
    - Who **can** you realistically serve in the first 3–5 years?
    - What user segment is within reach?
    - What percentage of your TAM is even accessible?

Then calculate your SAM:

- *Filtered user base or procedure count × Expected price or savings*

Add 2–3 key constraints you used to get here (e.g., "only Medicare-eligible patients," or "only outpatient use").

Step 3: SOM — Define the First Win

This is your executional North Star.

- Choose your **initial beachhead**: a region, system, or specific user group you could feasibly enter within 1–2 years.

- Based on outreach, pilot capacity, early partnerships, and budget:
    - What fraction of your SAM could you realistically reach and convert?
    - What is your best-case (but still realistic) **Year 1 revenue opportunity**?

Back it up with real-world constraints:

- Clinical capacity, team bandwidth, licensing timelines, funding limits, etc.

Write your SOM as a paragraph, not just a number. Show your reasoning, not your ambition.

Step 4: Map Who Pays vs. Who Benefits

Create a simple table with four columns:

- **Stakeholder**
- **Role (User / Purchaser / Payer / Influencer)**
- **What They Gain**
- **What They Spend / Approve**

For example:

Stakeholder	Role	What They Gain	What They Pay or Approve
Hospitalist	User	Saves time on discharge notes	Nothing (not a purchaser)
Clinical IT Manager	Purchaser	Improved system compliance	Approves software spend
Medicare (CMS)	Payer	Fewer readmissions	Approves DRG / bundled codes
Patient	Beneficiary	Safer discharge instructions	No direct payment

This will become a core tool in your fundraising and messaging strategy. It shows not just your value proposition, but who needs to believe in it and who needs to write the check.

Step 5: Check for Alignment

Once your table is built, answer:

- Is the person who benefits **also** the person who pays?
- If not, how will you create incentives to drive adoption anyway?
- Can you justify pricing in terms of cost savings, revenue generation, or competitive advantage?

Write a 1–2 paragraph reflection that anticipates friction, and suggests how you'll handle it.

Optional Stretch Goal: Visual Slide Draft

Using your TAM/SAM/SOM calculations and stakeholder map, sketch a first draft of your **market slide** for a pitch.

Include:

- 3 market size numbers (TAM / SAM / SOM)
- Key assumptions (with sources)
- Who pays vs. who benefits
- One sentence on early traction or planned go-to-market strategy

This is your chance to turn all of the research into a compelling story investors or funders can understand in 30 seconds.

Anchor This

You now have a first-pass market strategy rooted in something real: not hand-waving.

These numbers will evolve. Your SOM may shrink or grow. Your buyer profile may shift. That's normal. But the discipline of honest, grounded estimation will protect you from the single most common cause of early-stage failure: building something impressive that no one is willing (or authorized) to pay for.

TAM tells you it's worth dreaming.
SAM tells you where you can start.
SOM tells you how not to drown.

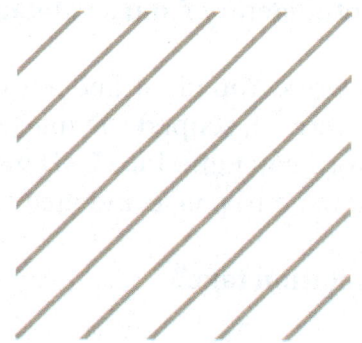

Part IV

From Idea to Form

CHAPTER 11: TRANSLATING NEEDS INTO CONCEPTS

How to brainstorm intelligently from a validated need

You've done the hard work. You identified a friction point that's real. You mapped the terrain, scoped the market, and validated the stakeholders. But before you jump into CAD models, wireframes, or circuit boards, you need to pause and answer a new question:

What form might a solution take?

This is the moment where good innovators go wrong. Some fall in love with their first idea. Others jump too quickly from problem to prototype without ever creating a full field of options. Both paths are dangerous. The purpose of this chapter is to help you avoid those traps and to show you how to generate, structure, and refine solution concepts in a way that leads to real-world impact.

STEP 1: START WITH THE RIGHT FRAMING

Ideation starts not with a whiteboard, but with **the right prompt**.

You are not asking, "What's the coolest thing we could build?" You are asking, **"What kinds of solutions could meaningfully address this validated need, given the constraints we've mapped?"**

The best ideation sessions begin with a **solution-independent need statement** and a clear articulation of the:

- End users
- Use environment
- Workflow implications

- Regulatory class (if known)

- Desired clinical or operational outcome

Your prompt should restate the need but avoid hinting at any specific technology or form factor.

Good: "Let's explore all possible ways to reduce line infections in patients with central venous catheters."
Bad: "How do we build a better catheter?"

Start wide. Specificity comes later.

STEP 2: BUILD THE RIGHT IDEATION ENVIRONMENT

Your brainstorming process should be:

- **Cross-functional:** Include voices from engineering, clinical practice, business, design, and even patient experience.

- **Low-ego:** Emphasize that no idea is final and nothing is too "wild" for early discussion.

- **Facilitated:** Choose someone to steer the session who has enough domain knowledge to push back but not so much that they dominate the thinking.

- **Time-bound:** Keep it under 90 minutes. Attention wanes after that.

Props help. Bring anatomical models, failed products, toy parts, whiteboard sketches, or anything tactile.

※ **BMTT Tip:** Assign a scribe. Every sticky note, quote, sketch, or napkin idea must be captured and archived. You'll revisit them more than you expect.

STEP 3: BRAINSTORM, DON'T BUILD

This is not the time to "solve the problem." Your goal is to generate **many distinct types of solutions**, not refine any single one. Encourage:

- Quantity over polish

- Tangents that explore adjacent ideas

- Sketching, building, or gesturing over lengthy explanation

- Unfiltered contributions: reserve evaluation for later

You're not looking for the perfect solution. You're looking to **understand the solution space**.

STEP 4: ORGANIZE AND CLUSTER IDEAS

Once you've got dozens (if not hundreds) of ideas on the wall, it's time to impose structure. Start clustering ideas using one or more organizing principles:

- **Mechanism** (chemical, mechanical, software-based, procedural)

- **Mode of use** (preventive vs. reactive, continuous vs. episodic)

- **Setting** (hospital, outpatient, home care)

- **User** (nurse, patient, surgeon, administrator)

- **Complexity** (off-the-shelf mod, new device, full platform)

You can also group ideas into:

- **Digital** (apps, software layers, AI tools)

- **Mechanical** (devices, surgical tools, physical components)

- **Workflow-based** (new protocols, checklists, automation)

- **Systemic** (policy, training, infrastructure changes)

※ **BMTT Tip:** Many of our most successful ventures begin with "boring" workflow innovations. Don't overlook system-level interventions just because they lack hardware or code.

STEP 5: BUILD A CONCEPT MAP

Visually document what you have. Put the need in the center. Surround it with clustered ideas. Subdivide clusters where needed. Use arrows or dotted lines to show synergy, sequence, or contradiction.

You can do this with:

- Post-it notes and a whiteboard

- Tools like Miro, Coggle, or MindMeister

- Custom-built slides for sharing with mentors or advisors

This map becomes a living document: one you'll return to repeatedly as you move from brainstorming to screening.

STEP 6: SCREEN FOR CONCEPT VALIDITY (NOT FEASIBILITY… YET)

At this point, you are not choosing your "final" solution. You are selecting a **small set of ideas worth exploring further**. That means checking each idea against:

- The original need statement

- The "must-have" criteria (outcome change, usability, clinical fit)

- Workflow and stakeholder compatibility

- Basic plausibility (not "can we build this?" but "can this work?")

This is not where you worry about IP, manufacturability, or regulatory classification in detail, but if something clearly violates those boundaries (e.g., impossible materials, banned procedures), remove it.

Optional Step: If you're uncertain which ideas are strongest, take a short list of 5–10 options and run quick feedback interviews with your clinical or operational advisors. Ask: "Which of these directions feels most promising or exciting and why?"

STEP 7: BIAS-CHECK THE RESULTS

Before moving forward, zoom out.

Ask:

- Are we unintentionally favoring one type of solution (e.g., devices over workflow changes)?

- Are we overlooking digital interventions?

- Are we assuming patients will behave perfectly?

- Did we overly rely on any single participant's idea?

If needed, run a second ideation round specifically to explore underrepresented categories. It's better to overbuild your idea map now than regret a blind spot later.

STEP 8: PATENT SCANNING

Before you move forward with prototyping, pause and make sure you are not chasing something that has already been done. Markets can be mapped with products, competitors, and user needs, but patents tell a different story: what has been protected on paper, whether or not it ever made it to market.

A patent scan is not about becoming a lawyer, it is about protecting your time. Think of patents as "invisible competitors." They reveal ideas that others thought were promising enough to protect, even if they never got off the ground. Sometimes you will find patents on concepts almost identical to what you planned. Other times you will find spaces that have been left untouched. Both outcomes are valuable.

This step matters no matter which camp you fall into. **If you are following this book in a problem-first way**, patent scans help you expand your sense of the landscape: where others have tried, where gaps remain, and where brainstorming energy should be focused. **If you came here with a pre-formed idea, perhaps from your own clinical or personal experience**, a patent scan is even more critical. It tells you quickly whether you are reinventing the wheel and helps confirm that your idea can be patentable before you spend untold hours developing it.

Why It Matters Now

- **Avoid Reinventing the Wheel:** If your idea is already patented and actively held, you need to know before you start building.

- **Spot Gaps in the Landscape:** Clusters of patents can point to crowded areas, while unexplored zones may show opportunity.

- **Identify Potential Competitors and Partners:** Companies with multiple patents may block you or may eventually partner with you.

- **Gather Design Insights:** Patents often contain detailed methods and diagrams that can spark refinements or pivots.

How Deep Do You Need to Go?

At this stage, you do not need an attorney's full Freedom to Operate opinion (that comes in a later chapter). What you need is a **basic scan** that answers three questions:

- Has someone already tried this?

- What approaches dominate the space?

- Are there clear gaps?

Free tools like **Google Patents** or the **USPTO database** are enough. Even thirty minutes of searching is often enough to show you whether you are exploring fresh ground or heading into a crowded field.

Key Point: Patent scanning is not just another layer of market research. It is a safeguard before prototyping and a way to ensure your energy goes toward something novel. Whether you are starting from a defined problem or a pre-formed idea, running a basic patent scan now saves you from wasted time later.

CASE REFLECTION: THE TRAP OF THE "OBVIOUS FIX"

Sana's team had locked in their need: reduce missed medication doses for elderly patients with chronic conditions. They'd interviewed dozens of caregivers, heard about pill confusion and memory lapses, and mapped workflows across home health visits, pharmacy drop-offs, and daily routines.

Their first idea came fast: a smart pillbox.

"It can beep, track doses, and sync with a caregiver app," said David, sketching it in the margins of his notebook.

Everyone nodded. It was familiar. Tidy. Buildable.

So they ran with it.

The Early Blind Spot

By week three, they had CAD mockups, a parts list, and an early Figma wireframe. But the further they got, the more friction they hit.

• Most of their target users had no Wi-Fi and didn't trust connected devices.

• Many lived alone and didn't notice beeps (or couldn't hear them).

• Caregivers weren't checking apps; they were too overwhelmed.

"It's a cool product," one nurse said during a feedback call. "But it won't get used."

Reframing the Prompt

At a mentor's urging, they went back to the whiteboard and rewrote their prompt:

"We are exploring ways to reduce missed oral medication doses in elderly patients, improving treatment adherence in home settings with low-tech users and minimal reimbursement coverage."

Then they re-ran their ideation session but this time with a geriatric nurse and a pharmacy tech in the room.

Ideas poured out that no one had considered before:

- A visual sticker system tied to meal times

- Pre-call scripts for pharmacy techs to confirm patient understanding

- A medication calendar printed and mailed weekly

- A low-cost blister pack redesign with punch-through tracking

The Breakthrough

One idea stuck: a *reversible color cue* on daily pill packs that turned white once pushed through, letting patients (and caregivers) instantly see what had been missed.

No Wi-Fi. No apps. No alerts. Just feedback.

"This could work anywhere," Sana said.
"And we don't have to fight user behavior. We meet it."

They hadn't abandoned the smart pillbox. But it was no longer the assumption. It was one of many.

FINAL THOUGHTS: IDEATION IS A PROCESS, NOT A PERFORMANCE

This phase is not about showing off your creativity. It's about building a structured, inclusive process to explore all the ways a validated need might be addressed.

Many early-stage founders treat ideation like a spark: one that either hits or it doesn't. But the best student innovators understand that real ideation is a discipline. You revisit it, test it, challenge it, and add to it as you learn more.

In the next chapter, we'll help you move from broad categories to early design exploration. We'll show you how to begin shaping these ideas with **real-world constraints in mind**: regulatory limits, clinical fit, user training, technical feasibility, and more.

But before you do that, make sure you've looked **wide** enough. If your only solution is your first idea, you're not building, you're gambling.

CHALLENGE: Generating with Discipline, Not Desire

You now hold a real, validated clinical need: one with a defined population, an observable outcome, a mapped landscape, and real-world terrain.

That's rare. It's earned. And it changes what comes next.

This challenge is not about dreaming big. It's about dreaming responsibly. Your goal is to generate a rich, diverse, and structured set of possible solution directions without drifting into solution bias, premature design, or unnecessary complexity.

The best ideas emerge not from your gut, but from disciplined creativity. Let's build toward that.

Step 1: Reframe Your Prompt

Begin by restating your need statement without hinting at a solution.

Use the structure:

- *"We are exploring ways to [solve the problem] in [target population] to [improve the outcome], taking into account [setting], [workflow], [regulatory class], and [user type]."*

Example:

- "We are exploring ways to reduce missed oral medication doses in elderly patients to improve treatment adherence, taking into account home settings, low-tech user profiles, and minimal reimbursement coverage."

This is your ideation anchor. You'll refer back to it throughout this exercise to make sure your ideas stay grounded.

Step 2: Generate 15–20 Distinct Solution Concepts

In a short, time-bound session (90 minutes or less), generate a wide range of potential concepts.
Use a mix of:

- Mechanical tools or devices

- Digital platforms or software

- Workflow protocols or systemic changes

- Preventative vs. reactive models

- Single-user vs. multi-user interfaces

Each concept should be **described in 1–2 sentences**: just enough to explain its mechanism or core approach.

Make sure at least 3 of your ideas:

- Don't involve building a new product

- Could be executed within existing infrastructure

- Are system-level or workflow-based

You're not picking your favorite. You're mapping your field.

Step 3: Cluster and Map Your Ideas

Create a **concept map** or cluster diagram that groups your 15–20 ideas by:

- Mechanism (mechanical, digital, procedural)
- User (nurse, patient, admin, etc.)
- Setting (hospital, outpatient, home)
- Novelty (incremental, new-but-grounded, blue-sky)

Use arrows or visual notes to connect ideas that:

- Could combine
- Contradict each other
- Address different stages of the workflow

This visual will help you zoom out and see your idea terrain: not just a single path through it.

Step 4: Screen for Validity

Narrow your 15–20 ideas down to 5 **high-potential directions**. For each one, briefly assess:

- Does it clearly connect to the original need statement?

- Does it align with workflow and stakeholder realities?

- Is it **plausible** (not "easy to build," but "likely to function")?

Write a 2–3 sentence justification for each of the 5, noting:

- What makes it promising?

- What major risks or unknowns remain?

This is not selection. It's prioritization for future refinement.

Step 5: Check for Bias and Blind Spots

Before you move on, take a hard look at your short list.

- Are you over-favoring digital tools because they're familiar?

- Are you avoiding workflow fixes because they're "less exciting"?

- Are you defaulting to hardware even when systemic change might be simpler?

If needed, revisit your long list and **force yourself to elevate one** idea that addresses a real need but wasn't getting enough attention.

This act of rebalancing will protect your solution space from premature closure.

Step 6: Scan the Patent Landscape (Lightly)

Before you finalize your short list, run a quick patent search on your top 5 solution directions. Use free tools like Google Patents or the USPTO database. This is not a legal exercise, it is a learning one. You are looking for:

- **Overlap:** Has someone already patented a very similar concept?

- **Clusters:** Are there many filings in one direction, suggesting the space is crowded?

- **Gaps:** Do you see fewer patents around one of your concepts, suggesting more room for exploration?

For each of your 5 ideas, capture a brief note:

- Did you find patents that look close?

- If so, what does that mean for your direction (pivot, differentiate, or keep as inspiration)?

- If not, what might that signal about opportunity or novelty?

This should take no more than 20–30 minutes per idea. The goal is not certainty. The goal is awareness.

Optional Stretch Goal: External Input

Share your top 5 solution directions with:

- A clinician

- A non-clinical stakeholder (administrator, caregiver, etc.)

- A peer in your project team

Ask:

- Which direction feels most buildable?

- Which one seems most likely to be used?

- What's missing?

Capture feedback in bullet form. These insights will help steer the early design phase that follows.

Anchor This

Ideation is not about brilliance. It's about breadth. It's about discipline. And it's about building a solution space wide enough and structured enough that the right answer has a real shot at emerging.

If you've done this well, you now have a concept map, a set of strong contenders, and a few surprises that didn't exist in your head a week ago.

Don't chase the one that sounds best in a pitch. Chase the one that fits the need, the workflow, the user, and the reality you now understand better than most.

CHAPTER 12: DESIGNING FOR REAL CONSTRAINTS

Co-Creation, Early Form Factor Decisions, and Building with the System, Not Against It

It's easy to get swept up in the excitement of a new idea. You've generated a range of concepts, grouped them, debated them, and maybe even started sketching one that feels "right." But before you crack open CAD software, start sourcing components, or build out your app wireframe... pause.

This chapter is about taking your **first real step into the build phase** without walking straight into a trap.

Here's the trap: thinking that a promising concept, validated need, and good intentions are enough to ensure adoption. They're not. Most products fail not because they don't work, but because they don't *fit*. They collide with the realities of how care is delivered, how institutions adopt technology, how users behave under pressure, and how systems resist change.

This chapter helps you design **with constraints in mind** (technical, regulatory, clinical, behavioral, and operational), so you can build something that actually lives in the world you're trying to change.

1. DESIGNING WITH, NOT FOR: CO-CREATION WITH REAL USERS

One of the most common mistakes student innovators make is designing in isolation. They get attached to a concept early and bring in users only after something's already built: usually when it's too late, too expensive, or too emotionally painful to change direction.

At BMTT, we emphasize **co-creation**, not as a buzzword, but as a practical and protective approach to early design. Co-creation means folding users into the design process early enough to let their reality shape your direction.

And importantly, this phase serves as a **pseudo customer discovery campaign**. It's not full-blown interviewing yet, but it's deeply observational. You're not validating your market here; you're vetting the feasibility of your concept's earliest assumptions through *use-informed creativity*.

Co-Creation ≠ Letting Users Design for You

Co-creation does *not* mean handing your idea over to a physician or asking a nurse to be your product strategist. It means treating their input as real-world design constraints.

A respiratory therapist might teach you that a filter must be positioned where it can be removed single-handedly. A home caregiver might tell you they can't use a product if it doesn't fit in a pocket or zippered pouch. These parameters are critical in understanding what specific needs your solution would need to consider.

Your job isn't to solve *for* them. Your job is to learn *with* them, in a way that translates their environment into design logic.

How to Start Co-Creating (Safely and Practically)

You don't need a fancy workshop to do this. You just need a semi-formed idea and an open loop of feedback.

Ways to co-create early:

- Host a casual sketch session with end users, care staff, or even other students in adjacent fields

- Print a few screenshots or hand-sketch workflows and ask "Where does this go wrong?"

- Hand a stakeholder a foam cutout or mock device and ask "What would annoy you about this?"

- Play out the "day in the life" scenario with someone who would actually touch or manage the tool

Optional but powerful: Invite 2–3 roles into the same session—e.g., nurse, patient, tech—and observe how their expectations conflict. That friction is gold.

※ **BMTT Tip: Bring in frontline workers and patients sooner than you think.**

While senior physicians may have influence, frontline staff and patients have *daily exposure* to the real frictions you're trying to solve. They tend to be less guarded and more direct, and their insights are often more practically useful in shaping early prototypes. We've seen teams course-correct major assumptions based on a single comment from a hospital tech or caregiver.

Guarding Your IP: Don't Overshare in the Excitement

One real risk during early co-creation is **premature disclosure** of your intellectual property.

Especially if your concept involves a potentially patentable feature, **you need to be cautious about how much you share, when you share it, and with whom.** In many jurisdictions (including the U.S.), disclosing your invention publicly without a filed provisional patent can start a 12-month countdown, or worse, immediately disqualify you from international patent protection.

Safe practices for early-stage co-creation:

- Share problems, not solutions: Focus discussions on friction points and conceptual feedback, not exact technical mechanisms.

- If you show a design, keep it **broad** and **labeled as a rough idea.**

- Keep a log: Record dates of idea generation and stakeholder feedback (this helps establish inventorship and protect timelines).

- If you're farther along and need to share more details, use a **basic non-disclosure agreement (NDA)**: especially with partners, not test users.

- Consider filing a **provisional patent application** through your university tech transfer office if your idea is mature enough.

✳ **BMTT Tip:** Don't let fear of disclosure paralyze your process. But don't be naïve either. Early-stage innovation is about finding balance: sharing enough to learn while keeping your invention protected.

---

2. CHOOSING THE RIGHT FORM FACTOR EARLY

The form your solution takes will shape everything that follows (from testing, to pricing, to regulation, to user perception).

Too often, student teams focus on functionality and delay decisions around **form factor**, assuming they'll "figure it out later." That's backwards. Form factor isn't decorative; it's strategic.

Common Form Factor Options:

- **Physical object/device**: Wearables, surgical tools, handheld diagnostics, etc.

- **Embedded device**: Something implanted or integrated into another system

- **Digital-only solution**: Apps, dashboards, workflow automation tools

- **Service/process protocol**: A new training regimen, scheduling tool, or procedure

- **Hybrid**: Devices with software layers, kits, or supporting systems

Each form carries different implications:

- **Physical** ⇨ regulatory complexity, sterilization, manufacturing concerns

- **Digital** ⇨ software validation, HIPAA compliance, interoperability

- **System/protocol** ⇨ buy-in, training, workflow modification

Questions to guide early decisions:

- Who handles it?

- Where is it stored?

- How is it maintained?

- How many times is it used?

- Can it fail silently? (If yes, higher design burden)

- What's the first thing a user does when it breaks?

Start answering these now, even if you're still working with cardboard and glue.

3. DESIGN WITH REGULATORY AND WORKFLOW BARRIERS IN MIND

You don't need to be a regulatory expert, but you do need to design as if your product will live inside a system that *resists* change.

That means:

- **Don't build a device that introduces an unvalidated step into a regulated workflow** (e.g., new dosing protocols, extra blood draws, new diagnostic categories)

- **Don't create tools that require surgical retraining unless you have a clear education plan**

- **Don't require IT integration unless you've validated that access is possible**

Early product choices should **reduce the risk profile**, not escalate it.

✗ **Example:** If a wound care patch requires refrigeration, you've now added a new storage standard to the clinic. Will they adopt that?

✗ **Example:** If your app requires pulling data from the EMR, you've now involved the IT department; does that clinic even have dev support?

> ⁂ **BMTT Tip:** One of the smartest early design moves you can make is to **repurpose or layer onto existing systems**. Design a surgical aid that works with tools already in the tray. Build a digital interface that mimics the format of EMR screens. Reduce new behaviors by extending familiar ones.

4. DESIGN THINKING (USED THE RIGHT WAY)

"Design thinking" has been buzzwordified beyond recognition. But at its core, it's a practical, repeatable framework for shaping ideas around **user constraints**, **rapid testing**, and **purposeful iteration**.

Here's how to actually use it without the jargon:

▦ Step 1: Empathize

Spend time with the people you're building for. Watch them work. Watch them struggle. Don't just ask what they want, ask what they *do* when something doesn't work.

▦ Step 2: Define

Use that observation to reframe your idea. "This isn't a tool for reducing error; it's a tool for reducing the *likelihood* that someone will forget to scan a barcode when rushed."

▦ Step 3: Ideate

Go wide with solutions but tie every idea to a real user behavior or constraint. Don't brainstorm in a vacuum.

▦ Step 4: Prototype

Make fast, disposable versions of your concept. Focus on *function*, not fidelity.

⚏ Step 5: Test

Put it in someone's hands. Watch what happens. Don't explain it: observe it.

❋ The BMTT Twist:
We encourage our teams to use design thinking *not* to refine a single solution, but to pressure-test multiple directions at once. You don't need to "pick a lane" until you've seen how users react to at least two very different approaches.

5. Building in Constraints Makes Your Life Easier Later

This chapter isn't about killing creativity. It's about protecting your time.

Student teams that design inside real-world constraints:

- Get better feedback during early testing

- Move through validation faster

- Attract more credible partnerships and funding

- Waste less money on MVPs that don't fit the system

If your early sketches, prototypes, or simulations already account for user behavior, institutional friction, and basic clinical logic, you're not behind. You're way ahead.

CASE REFLECTION: WHEN THE CAD MEETS THE CLINIC

Marco's team had a slick concept: a handheld ultrasound accessory that clips onto a smartphone to enable rapid bedside imaging in rural clinics. They'd validated the need, sketched CAD mockups, and built a clickable app prototype. Everything looked on track.

But their first co-creation session told a different story.

"Cool tech, but it won't survive in our workflow," said a community nurse, handling their foam mock device with gloved hands.
"It fogs up. We don't charge phones. And what happens if it falls? We don't have replacements lying around."

Suddenly, the constraints they'd mapped fit into real, lived experience.

Co-Creation Resets Everything
It turned out the nurse's biggest worry wasn't image quality; it was **durability and dependability**. In clinics with intermittent power and crowded exam rooms, a device that relies on smartphone charging and fragile alignment simply wouldn't be used.

Form Factor Shift
They sketched a new concept on a napkin in the aisle of the clinic. Instead of a fragile smartphone clip, what if it was a **wrap-around probe with built-in controls and a disposable sheath**? No phone needed, recharges fast via solar USB, and no fear of damage if dropped.

Constrained Creativity
Using their five top constraints (**budget under $250, waterproof casing, glove-friendly buttons, standalone battery, and no EMR integration required**), they mapped three competing form ideas from earlier brainstorming:

1. **Smartphone clip** (original)

1. **Standalone probe with screen**

1. **Workflow-driven central imaging station**

They evaluated which would:

- Survive in rainy, low-power settings
- Be intuitive for minimally trained users
- Cost less than $250

The standalone probe won out. It checked all real-world constraints and even earned the nurse's "I'd punch a button for this" stamp of approval.

Takeaway
Without co-creation and constraint-checking, the team was building for *potential*, not *context*. What started as promising technology almost failed because it clashed with real workflows. But by designing WITH constraints, not against them, they created something users actually wanted and could use.

FINAL NOTE: CONSTRAINT IS A CREATIVE TOOL

Design isn't about blue-sky imagination. It's about building something real for someone specific, in a context that resists novelty. The best innovators aren't the ones with the wildest ideas. They're the ones who figure out how to make a **new idea fit**.

In the next chapter, we'll show you how to move from those early decisions into tangible prototypes. You'll learn how to build fast, test smart, and know whether your early concepts are pointing you in the right direction.

Spoiler: most prototypes are wrong the first time. That's the point.

CHALLENGE: Co-Create With Constraint

You've done something few early innovators ever do: you've validated a real need, mapped the terrain it lives in, explored multiple solution paths, and resisted the urge to build too early.

Now, with your highest-potential concept in hand, it's time to take the first real design steps but not alone, and not in a vacuum.

This challenge is about shaping your idea inside the boundaries of the real world. It's your first act of translation: from concept to candidate.

Step 1: Run a Co-Creation Session with 1–3 Stakeholders

Choose at least one stakeholder from your earlier mapping (ideally two or more, including one frontline worker or patient). Set up a short, structured session (30–45 minutes is enough).

Bring:

- A rough sketch, wireframe, diagram, or mock-up (paper, foam, even slides)

- A walkthrough scenario: "This is what your day might look like with this product in use."

- A notepad and an open mind

Prompt the session with:

- "What would annoy you about this?"

- "What would break first in your environment?"

- "What would make this go unused a week after launch?"

- How much time would it take you (or your team) to understand and feel comfortable using this in your daily workflow?

Capture everything: words, reactions, hesitation, confusion, workarounds.

Deliverable: Write a 1–2 paragraph design reflection. What did you expect to hear and what surprised you?

Step 2: Define Your Early Form Factor

Your solution can't stay ambiguous. You need to commit to a **first-form assumption**.

Will this be:

- A **physical product**?

- A **digital interface or tool**?

- A **workflow or system change**?

- A **hybrid**?

Now answer the following, as specifically as possible:

- Where will this product live when not in use?

- Who will carry it, clean it, recharge it, explain it?

- How will it be taught, updated, and monitored?

Deliverable: Write 3–5 bullets describing your form factor and the immediate implications it creates for development, workflow, and regulation.

Step 3: List the Top 5 System Constraints That Shape Your Design

You don't need a regulatory expert or a manufacturing plan yet, but you do need constraint-awareness.

Based on your landscape work, stakeholder input, and initial design sketching, list five real-world constraints your product must accommodate. These should be **non-negotiable.**

Examples:

- Must avoid EMR integration
- Must not require refrigeration
- Must be usable with gloved hands
- Must be under $100 per unit
- Must not interfere with medication schedules

These constraints will sharpen your design, not limit it.

Deliverable: List your five top constraints and describe (in one sentence each) how they impact your design direction.

Step 4: Map Three Competing Concepts Inside These Constraints

Go back to your idea map. Select **three distinct concepts** from earlier brainstorming that still seem promising.

Now pressure-test them:

- Could each work within the constraints you just defined?

- Which is **most resilient** to system complexity?

- Which **requires the least behavioral change**?

- Which one would a user adopt tomorrow if it existed?

- Which concept best balances desirability, feasbility, and viability when viewed through your stakeholder ecosystem?

- Where are the largest unkowns that would need early prototyping or customer discovery to de-risk the path forward?

Deliverable: Write a short paragraph on each idea. Rank them from most to least aligned with real-world fit: not excitement, but feasibility.

Anchor This

This chapter's work separates builders from dreamers.

You've now taken the first step from idea to artifact. Not by chasing novelty, but by designing within real-world parameters: workflows, behaviors, storage closets, operating rooms, and end users who are too busy to adopt bad design.

You've sketched in constraint, co-created in context, and chosen a form you can actually build. From here, we move into prototyping not as guesswork, but as grounded experimentation.

CHAPTER 13: PROTOTYPING LIKE A BUILDER

Tools for scrappy prototyping (digital, physical, hybrid)
No-code tools, lab access, 3D printing hacks, and collaborative workarounds

You're ready to build. Or at least, to build *something*.

But here's where many student teams freeze:
"I don't have a lab."
"I don't know how to code."
"I'm not a designer."
"I don't have $10,000."

Here's the truth: none of those are valid blockers at this stage.

Prototyping isn't about perfection. It's about *proof*. Can you show that one small part of your idea works well enough to move forward? That's the bar. And if you can hit it with paper, pipe cleaners, Figma, or foam, do it.

This chapter will teach you to:

- Build fast and cheap to answer key design questions

- Use tools that are free, accessible, and beginner-friendly

- Work across digital, physical, and hybrid mediums

- Translate early sketches into testable forms

We'll also introduce some of the most effective free tools you can start using today, many of which we use inside BMTT.

Let's build scrappily, strategically, and with purpose.

1. START WITH THE RIGHT QUESTION

Every prototype should exist for a reason. Before you open a design tool or buy a roll of filament, write down one sentence:

"This prototype is meant to help us understand whether _____."

That blank might be:

- Users understand how to interact with the layout
- The component fits on an exam table
- A fluid system will create enough pressure
- An alert system gets noticed within 5 seconds
- The part can be gripped one-handed

You are not trying to validate the whole product. You're isolating one **critical path question** and building the absolute minimum necessary to answer it.

※ **BMTT Tip:** Think in "functional blocks." What are the main elements of your solution (mechanical function, user input, chemical reaction, power source, display)? Only build what helps you test the riskiest one first.

2. CHOOSE THE RIGHT PROTOTYPING TOOLS

You don't need to start with complex CAD models or PCB design. Choose tools based on the problem you're solving and what you're building (physical, digital, or hybrid).

Lean Prototyping Tools

Used for cheap modeling with materials like foam core, modeling clay, zip ties, and even tool packaging.

✖ Example: One of our first intraorbital device prototypes was made from a cut piece of plastic packaging and a dab of modeling clay.

3D Design & Rendering Tools

Use these to visualize or simulate early physical prototypes:

Tool	Free Plan?	Best For	Example Use Case
Shapr3D	✓ Free	CAD on iPad or PC	*Cervical Collar* – ergonomic neck brace design
Fusion 360	✓ Free for students	Advanced CAD with simulation	*Drug Delivery Device* – simulate catheter fluid dynamics
Tinkercad	✓ Free	Simple 3D + electronics	*POTS Screening Tool* – wearable heart rate tracker

UI/UX Wireframing and App Prototyping

Perfect for digital health tools, AI-driven apps, and patient tracking systems.

Tool	Free Plan?	Best For	Example Use Case
Figma	✓ Free	Interactive wireframing	*AI Wound Alert System* – mobile app for wound tracking

Tool	Free Plan?	Best For	Example Use Case
Framer	✓ Free	Clickable, high-fidelity mockups	Smartwatch Alerts – simulate missed medication warnings

No-Code / Low-Code MVP Builders

For semi-functional digital tools without coding:

Tool	Free Plan?	Best For	Example Use Case
Webflow	✓ Free	Website mockups	*Dialysis Scheduling Optimizer*
Bubble	✓ Free	No-code web apps	*AI Pill Timing Dashboard*
Thunkable	✓ Free	Mobile apps	*AI Physical Therapy App*

❊ **BMTT Tip:** If you're testing a clinical flow or user interface, focus first on *perception*. Can the user understand it without a manual? If not, you're not ready to build more.

---

3. BUILD WITH READILY AVAILABLE MATERIALS

Don't wait for lab access. Don't let yourself believe you need a sterile room or an NIH grant.

Build your *first* prototype with what you can get in 48 hours:

- Foam core, tape, and bendable wire

- 3D-printed PLA using public library printers

- Syringes, tubing, and hardware-store clamps

- Old device parts from medical surplus websites

- Modeling clay, tool packaging, or parts from broken toys

Borrow, repurpose, and improvise. You're not building a product; you're building *confidence in a concept.*

4. MAKE INCOMPLETE PROTOTYPES TESTABLE

Your prototype does not need to work fully to be useful. In fact, most early prototypes *shouldn't*.

If your system involves multiple components, test just one:

- Build the *software wireframe* and simulate hardware interaction manually

- Record a demo video to simulate sensor behavior or alerts

If you're building an implantable device but can't simulate long-term use, test for:

- Fit and comfort

- Handheld usability

- Cleaning/reprocessing logic

If you're prototyping software but not connected to real data, test:

- The *workflow* (does it make sense?)

- Button placement and screen transitions

- Alerts and nudges

✲ **BMTT Tip:** Label your prototype when testing: "This is a non-functional model to assess size, grip, and visual design." - It keeps users focused on what you're trying to learn, not what you haven't built yet.

5. DON'T OVER-PROTOTYPE

This can't be said enough: **you can prototype too much.**

If you build a fully functional MVP before confirming that the workflow fits or the user actually wants it, you've wasted time and risked demoralizing your team. This is especially common in technical teams that "build for the sake of building."

Remember: prototyping is a tool to **reduce risk**, not a trophy.

FINAL THOUGHTS: PROTOTYPE WITH PURPOSE

Prototyping is an exploratory tool. It's how you retire key risks, surface unexpected flaws, and iterate ideas without major cost.

Think scrappy. Think fast. Think directional. You are *not* proving final performance. You are **asking better questions, faster.**

In the next chapter, we'll zoom out to show how your prototype, and the learnings it generates, fit into a broader testing framework: usability, human factors, stakeholder feedback, and even simulated clinical flow.

But for now, build smart. Test often. And remember, duct tape and Figma are still innovation tools if you know how to use them.

CHALLENGE: Build to Learn, Not to Impress

You've crossed the line from planner to builder.

You now hold a real concept, rooted in clinical need, informed by system fit, and shaped by constraint. Your job is not to prove the concept right: it's to learn what's wrong, quickly and cheaply, before those mistakes get expensive.

This challenge helps you prototype with purpose, pressure test what matters most, and develop scrappy confidence in your ability to build.

Step 1: Define the Prototype's Job

Every prototype must earn its keep. Before you start crafting, write one clear sentence:

"This prototype is meant to help us understand whether _____."

You're not trying to prove your product works. You're answering one critical question.

Examples:

- "Whether a patient can understand the alert system without training"

- "Whether the form factor is comfortable in a pocket for 8+ hours"

- "Whether the app's flow makes sense without verbal explanation"

Deliverable: Write your prototype's job statement in one sentence.

Step 2: Choose a Form and Tool Based on Function

Select a tool or method that will help you answer that question *without* overbuilding.

Choose from:

- Foam core, cardboard, or modeling clay
- Public or university 3D printers (PLA is fine)
- Figma, Framer, or PowerPoint for digital wireframes
- No-code tools (Bubble, Webflow, Thunkable)
- Combination prototypes (physical shell + digital slide simulation)

The goal is testability, not fidelity.

Deliverable: List your chosen materials/tools and the reasoning behind them.

Step 3: Build a Single-Risk Prototype

Focus on the **riskiest assumption**. This is the place where, if your idea fails, the whole concept collapses.

Examples:

- If comfort is the make-or-break feature
  ⇨ build a wear-testable dummy

- If digital clarity is essential
  ⇨ build a tappable Figma prototype

- If motion or mechanics matter
  ⇨ simulate the interaction, even manually

Deliverable: Create your prototype. Attach 1–3 photos, screen recordings, or slide mockups.

You do **not** need functionality. You need feedback.

Step 4: Label the Test and Run a 5-Minute User Walkthrough

Give your prototype a purpose label:

"This is a non-functional form model to test patient comfort during transport."

"This is a UI mockup to evaluate alert visibility under time pressure."

Put it in front of a real user such as a patient, clinician, caregiver, or adjacent role, and observe their reaction. You may explain setup but not function.

Ask:

- "What would confuse you here?"

- "What would annoy you about using this for 10+ hours?"
- "Where would this go wrong in your daily routine?"

Deliverable: Write a short reflection on what you learned. What worked? What didn't? What surprised you?

Step 5: Revise Based on What You Heard, Not What You Built

Even if it hurts, use that feedback to improve one thing. Add a label. Change a screen. Reshape a grip.

Do not justify the design: improve it.

Deliverable: Make a visible change to your prototype and summarize how that change addresses user feedback.

Anchor This

You are now a builder. Not in the sense of final products, but in the sense of forward motion.

You've taken a concept out of your head and into the world. That single act of making an idea tangible will teach you more than ten brainstorms ever could.

This wasn't about polish. It was about pressure-testing.

Part V

Testing the Right Things

CHAPTER 14: TESTING FOR VALUE AND USABILITY

From Concept to Functional MVP With the Right Kind of Feedback

Before we jump into broad customer discovery and stakeholder interviews, there's a quieter stage you need to pass through first.

This is the *refinement phase*. Not validation. Not scaling. Not testing for virality or adoption. Just building a usable version of your concept with enough real-world friction baked in to know it won't crumble the second it touches the system.

At this point, you don't have a finished product.
You have a **raw solution** (a mechanism, layout, process, or flow) that addresses a problem you've carefully framed.

Your goal in this chapter is to *refine it into a functional* MVP, one shaped by targeted testing with a **small, trusted circle of informed stakeholders.** Not because they represent the full market, but because they'll pressure-test your assumptions, usability, and practical fit.

This is your **pre-customer discovery proving ground.**

Why This Comes Before Customer Discovery

It's a common trap to bring a half-baked prototype to a full discovery interview. When that happens, you're not learning from the market, you're apologizing for your build.

Customer discovery is about listening. But to have a real conversation about your idea, you need **a thing worth reacting to**, something with basic form and functionality that people can *interact* with, not just imagine.

That's what this chapter gets you to:
A functional MVP with the right form factor, core functionality, and enough contextual testing to survive its first serious scrutiny.

START WITH SEMI-PRIVATE TESTING

You only need 1–3 informed users for this stage. These people:

Are directly affected by the problem

Know enough context to give grounded feedback

Won't sugarcoat usability issues

Are willing to work with you more than once

This is not about scale. It's about trust and tactical honesty.

⊠ Example profiles:

- A frontline nurse who sees the problem daily and can walk you through it
- A specialist who's willing to test multiple early form factors and tell you what's clunky
- A caregiver who can try your interface or device over several days and report friction

What You're Testing Now

You're not testing whether your startup is viable.
You're testing whether your **product idea is even functional in context.**

At this stage, you should be refining:

- Form factor: Does it physically or digitally fit?

- Workflow fit: Does it interrupt or integrate into the existing process?

- First-use clarity: Does the user know what to do without help?

- Barriers: Setup time, confusion, poor UX, accessibility

You *shouldn't* be worrying yet about:

- Market sizing

- Reimbursement or pricing

- Customer segmentation

- Institutional procurement

That comes next.

DON'T BUILD EVERYTHING, JUST THE RISKY PARTS

You don't need to build the full app, the whole device, or a finished system. You just need to **retire the highest risks.**

Ask:

- What is most likely to break, confuse, or fail?

- What are the top 1–2 user interactions that must be intuitive?

- What feature, if wrong, would tank the whole solution?

Build only what helps answer those questions. Everything else can be placeholder.

Pair Prototypes With Workflow Observations

One of the best ways to test usability is **in context**.

Try this:

- Shadow the stakeholder during their normal task (rounding, setup, logging, handoff)
- Ask when and how your product might appear
- Introduce your prototype *at that exact moment*
- Watch how they use it without instruction

This isn't formal testing. This is **casual, contextual observation**. But it reveals more than feedback ever will.

✷ **BMTT Tip:** When someone hesitates or fumbles, don't explain. Wait. The silence will teach you more than your script.

DON'T CONFUSE ENCOURAGEMENT WITH ADOPTION

As a student or early-career builder, people will *encourage* you. They'll admire your effort. They'll smile and say, "This is neat." They might even let you shadow them or show your prototype to their team.

That's not the same as wanting to use it tomorrow.

Ask:

- "Would you replace your current workflow with this right now?"

- "What would have to be true for this to be used today?"

- "What is the first thing you'd change or remove from this?"

USE LOW-FIDELITY BUILDS TO TEST HIGH-STAKES INTERACTIONS

You can test function *without* function.

- Paper wireframes or clickable demos for software

- Foam mockups or 3D prints for physical devices

- Simulated alerts or workflows using slide decks and voiceovers

These "throwaway" prototypes reduce time and pressure, and help users focus on *interaction*, not polish.

※ **BMTT Tip:** Every time you level up fidelity, ask what new feedback you're trying to extract. If it's still surface-level, your prototype is probably *too complete*.

Know When to Strip Down or Pivot

This stage is your best shot at preventing overbuilding.

- Is a feature consistently skipped, misunderstood, or resented? Cut it.

- Are users making workarounds to avoid part of your flow? Change it.

- Are testers more excited about one narrow use case? Pivot to that.

Your MVP should **do less, better.**

Sidebar: Use Your Student Status to Get In, Then Listen Like a Builder

As a student, you have unmatched access:

- Shadowing is easier to secure

- Clinicians are more receptive to mentoring

- Institutions view you as non-threatening

But that charm can blur the feedback. People will soften criticism. They'll project support.

Your job: Use student access to test hard truths, not to collect praise.

Let them like *you*, but challenge *your product*.

Sidebar: What "Evidence of Use" Really Means

- A nurse saying "That's cool" ≠ Willingness to log it 4 times per shift

- A physician who compliments your alert ≠ Integration into their EHR workflow

- A patient who smiles ≠ Consistent, unsupervised use

Look for **repetition**, not reaction.

True "evidence of use" is:

- A prototype that gets used the way you intended
- A task completed faster, easier, or with fewer steps
- A stakeholder who asks when they can try it again

CASE REFLECTION: THE MVP MIRAGE

By week nine, Kara's team had something real: a sleek, clickable mockup of their patient-side symptom logging app. Clean interface. Smooth transitions. Everyone on the team agreed that it *felt* ready.

They booked a session with a nurse who had helped them earlier in the process.

"This looks great," she said, scrolling politely. Then: "But when would the patient ever be the one entering this?"

Kara blinked. "What do you mean?"

"In our setting, it'd be the caregiver. Or a nurse aide. The patient can't use a screen: some can't speak, most can't move their arms. This would just... sit there."

Silence.

The team had built for the wrong user.

And it wasn't a minor oversight. The symptom flow was intuitive but only if you had full motor control and a private hospital room. Neither applied here.

They were missing the necessary context needed to make a logging app that would actually solve the problem.

The next week, they tried again. This time, they shadowed during morning rounds, introduced the prototype at the bedside, and watched who reached for it.

It wasn't the patient.

It was the night nurse, looking to log handoff notes, squinting at the screen.

"Why can't it just print a summary?" she asked. "That's what we file."

So they scrapped the symptom entry altogether. Pivoted to a voice-activated summary tool for overnight nurse notes. Same app framework, new workflow.

The UI stayed. The purpose changed.

Takeaway:
Good feedback doesn't just improve your build. It can *redirect the entire product.* **You don't test MVPs to validate what you made; you test to learn what should've been made in the first place.**

FINAL THOUGHT: YOUR MVP ISN'T READY FOR THE WORLD YET, BUT IT'S CLOSE

The goal of this phase is to get to your **first real MVP**: something that can anchor meaningful customer discovery conversations.

It's still raw. It's still incomplete. But it should:

- Work in real contexts
- Fit within known constraints
- Survive honest feedback
- Represent what your startup *could* become

Now you're ready to go wider.

In the next part, we'll teach you how to run high-impact discovery campaigns that don't waste time and don't rely on polished pitches.

You're not proving yourself anymore. You're **proving the problem**.

CHALLENGE: Earn Your MVP

Now comes the most important work: shaping that concept into a **functional MVP**. This should not be a final product, but a real-world artifact that someone can actually use, critique, and learn from.

This isn't about polish. It's about *proof of usability*. And it starts by leaving your comfort zone.

Step 1: Define What "Usable" Means for Your MVP

Your MVP must pass a simple test: it can be picked up, understood, and used (however partially) by a real stakeholder in a real or simulated setting.

Ask:

- What *must* the user be able to do?
- What *can't* go wrong in early use?
- What part of the experience should feel intuitive?

Deliverable: In 3–5 bullets, define what "minimum usable" means for your concept.

Step 2: Identify Your Feedback Inner Circle

You only need 1–3 at this stage:

- Someone affected by the problem

- Someone honest enough to tell you what's broken

- Someone open to testing rough, incomplete builds

These aren't general users. They're **truth-tellers** with context.

Deliverable: List 2–3 names or roles you'll approach for this round. Note why each is valuable and what risk they'll help you test.

Step 3: Conduct a Contextual Walkthrough

Don't just demo your MVP: **embed it in context.**

Try:

- Having your stakeholder go through a "day in the life" scenario while your product enters naturally

- Watching them use it in a simulated task without instruction

- Running the interaction where the real friction occurs: at the bedside, in the lab, on the chart, or in the home

Deliverable: Write a brief summary of what happened. What surprised you? What failed? What felt clearer than expected?

Step 4: Pressure Test for Real Use

Encouragement is cheap. Adoption is hard.

Ask your testers:

- Would you replace your current method with this today?

- What would need to be true for this to feel usable now?

- What feature would you cut or change immediately?

You're not hunting for praise. You're hunting for friction.

Deliverable: List 3 quotes, reactions, or pain points from your test that you will act on in your next iteration.

Step 5: Refine Your MVP (by Cutting, Not Adding)

Simplify based on what broke. Most MVPs aren't missing features; they're overloaded.

Trim until:

- One critical interaction is intuitive

- One stakeholder can use it without help

- One context proves it belongs

Deliverable: List the 1–2 most meaningful changes you made after testing. Explain *why* you made them.

Anchor This

You've now crossed one of the hardest thresholds in early venture building: moving from prototype to product logic. You're no longer guessing. You're watching. Listening. Stripping away the wrong parts and sharpening what matters.

This MVP won't be your final version. But it will be your **first honest one**: the one you'll carry into the next phase.

Because in the next chapter, you will discover that the market won't care what you built. It'll care whether it solves a problem that people are already desperate to fix.

So build less. Observe more. And carry only what works.

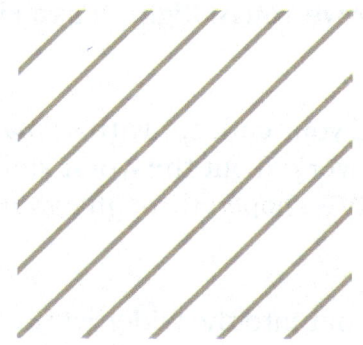

Part VI

Customer Discovery with Confidence

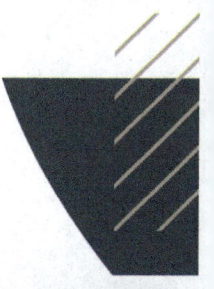

CHAPTER 15: CUSTOMER DISCOVERY WITH CONFIDENCE

The Goal Is Not to Prove You're Right; It's to Find Out Where You're Wrong

By now, you've tested your concept with a small circle of stakeholders. You've worked out the worst kinks. You have a raw but functional MVP, shaped through careful iteration and usability trials.

Now it's time to step out into the wider world but not to pitch, not to sell, not even to "get feedback" in the way most founders imagine.

This is **customer discovery**, and its primary goal is simple: **Try to invalidate your solution before the market does.**

Invalidation Is the Real Success

Too many founders treat customer discovery as a stage for praise. They show their prototype, ask leading questions, and subconsciously fish for validation. Every nod feels like a win. Every "I like this" becomes fuel.

But that's not discovery. That's delusion.

Your job now is to **hunt for reasons your idea might not work**:

- Where does it fail to fit?

- Who doesn't need it?

- What parts are pointless?

- Where are the real barriers?

This phase isn't about making your prototype look better. It's about sharpening your understanding of the problem, the context, and the stakes.

What You're Really Trying to Learn

When done right, customer discovery reveals:

- Who truly feels the problem (and who doesn't)

- How those people currently solve it
  (and how well that works)

- What alternatives they've already tried (and why they failed)

- What institutional, regulatory, or financial forces drive adoption

- What emotional or cultural barriers could still stop you

If your MVP survives this process, it's on solid footing.
If it doesn't, you're lucky to find out now.

HOW TO DESIGN DISCOVERY CONVERSATIONS THAT ACTUALLY WORK

Good customer discovery campaigns aren't random interviews. They're **structured learning systems**.

Start by building a plan that includes:

- Segmented outreach by role (e.g., physicians, nurses, lab managers, execs)

- Role-specific questions (see end of chapter for examples)

- A mix of interviews and structured survey tools (like Google Forms)

- A weekly rhythm of review and revision based on early responses

You're looking to run **campaigns**, not one-off chats.

Customize for the Role, Not Just the Institution

A surgeon and a floor nurse may work in the same hospital, but their pain points and priorities couldn't be more different.

Examples:

- **Surgeons** care about speed, precision, and patient outcomes in the OR. Ask them about procedural bottlenecks, tool reliability, and post-op risks.

- **Nurses** care about time allocation, data entry, patient comfort, and task switching. Ask them about interruptions, device annoyances, and team communication.

Every role touches the system differently. Your questions should reflect that.

※ **BMTT Tip:** Create separate Google Forms for each stakeholder category. Design role-specific versions of your discovery questions. This helps you compare patterns *within* a role and track insights *across* them.

Question Design: The Heart of Discovery

Don't ask: "Would you use this?"

That's a yes trap. Instead, ask:

- "What's the biggest challenge in your current workflow when doing X?"

- "How do you currently solve that?"

- "What's frustrating about your current solution?"

- "Have you tried any other options? Why did or didn't they work?"

Ask to explore problems, not validate solutions.

When you do ask about your prototype, frame it as a tool to test their mental model, not yours.

⊠ For example:
- "Does this help with [X]? What's missing?"

- "Would this be easier or harder than your current method?"

- "If this existed tomorrow, would it change anything about how you work?"

Optional Tip: *"What do you think?" is the weakest question in your toolkit.*
It invites politeness. Instead ask:

- "What would stop you from using this?"

- "Who would kill this in your department?"

201

- "When would you *not* use this?"

Power Dynamics: Feedback Without a Title

You might be a student. Or early in your career. Or a solo builder.

That means you're not showing up with a badge, a brand, or a title.

Use that to your advantage.

- Clinicians and execs often **lower their guard** when they know you're not selling something.

- Frame yourself as a **learner**, not a founder.

- Introduce your work as a project meant to explore pain points, not pitch solutions.

- Let your questions do the proving.

※ **BMTT Tip:**

Use phrases like:

- "I'm not trying to sell you anything; I am trying to understand how you experi**ence [X]**."

- "You are the expert here. I am just trying ot learn where we'd fa**ll short**."

- "If this isn't useful, I'd much rather **know now**."

MAKE DISCOVERY ITERATIVE

Customer discovery is not static. It's not a one-week sprint. It's a rolling campaign that evolves with every conversation.

Your rhythm should include:

- Weekly reviews of new data

- Real-time updates to your question sets

- Emergent theme tracking (i.e., "4 people mentioned setup time")

- Strategic branching ("Let's add more questions about discharge workflows")

You're not just getting feedback. You're building a **map of the ecosystem** you're trying to enter.

On Protecting IP (Without Hiding the Problem)

If you're validating a prototype, not just exploring problems, you need to protect your work.

That means:

- Keeping descriptions focused on **what** the solution addresses, not **how**

- Using NDAs for sensitive conversations (especially with execs or purchasers)

- Avoiding deep dives into technical specs during early feedback loops

You're not pitching your IP. You're testing your understanding of the problem.

Reframe the conversation:

- "Here's what this solution aims to accomplish. Does that matter to you?"

- "Here's the problem we're trying to solve. How do you handle that now?"

Let them confirm the **problem** before you reveal the **product**.

Sidebar: Two Types of Discovery (And How to Know Which You're Doing)

1. Pain Point Discovery

Use this when: You're still hunting for what matters
Focus: Open-ended questions, no mention of a specific product
Goal: Understand pain, context, and gaps in current solutions

❌ Example:
"What's your biggest bottleneck during a 12-hour shift?"
"What do you wish your current tools did better?"

2. Prototype Validation Discovery

Use this when: You have a working MVP
Focus: Problem confirmation and use-case alignment
Goal: See if your solution aligns with real workflows

❌ Example:
"If a tool helped you reduce [X], would it be useful? What's missing from this approach?"

Never blur the two. Each phase requires different questions, different framing, and different stakes.

Final Thought: The Best Feedback Is What You Didn't Expect

Great discovery campaigns don't just confirm what you thought; they break your assumptions.

They tell you:

- Who your product *isn't* for

- Which problems actually aren't worth solving

- Why your MVP is one feature too bloated, or one insight too shallow

They make you rethink, reframe, and reposition. And they give you the confidence to build forward without guessing.

You're not listening for praise. You're listening for friction. When you find it, lean in.

CUSTOMER DISCOVERY SCRIPT BUILDER

Overview:
This builder is designed to help you structure, iterate, and improve your customer discovery campaigns. It includes:

- A universal template with role-based customization

- Framing tips for every section

- A running example that evolves over multiple interviews

Use this worksheet before any discovery call, update it after each interview, and treat it as a living document that evolves with your learning.

Part 1: Define Your Goal

What are you trying to learn at this stage?

Choose one:

- ☐ Discover unmet pain points

- ☐ Validate a perceived problem

- ☐ Refine an MVP

- ☐ Understand stakeholder priorities

- ☐ De-risk a go-to-market assumption

⊠ **Example:**
"We want to understand how interventional cardiologists experience failure or recurrence after angioplasty, and whether our assumptions about procedural frustration hold true."

Part 2: Know Your Interviewee

Category	Notes
Stakeholder Type	e.g., Surgeon, Nurse, PA, Purchasing Manager
Role-specific Concerns	e.g., Time pressure, documentation burden, cost thresholds

Category	Notes
Decision Power	Influencer, recommender, decision-maker, end-user

Part 3: Build Your Core Questions

Organize your script into **four categories**:

1. Intro / Rapport Building

- "Can you tell me a bit about your role and what a typical day looks like?"

- "How long have you been working in [environment]?"

Purpose: Sets context. Establishes trust. Positions you as a learner.

2. Problem Discovery

- "What part of your workflow is most frustrating, especially during [procedure]?"

- "Have you had to create any workarounds in your process?"

- "Is there any tool or system you've stopped using because it didn't fit your workflow?"

Purpose: Explores friction, inefficiency, emotional cues.

3. Solution Fit (if MVP exists)

- "This is something we're exploring; it aims to [broad benefit], without going too deep into how it works. Based on your experience, would this change anything?"

- "If this tool had existed a year ago, what would have happened differently?"

- "What's missing?"

Purpose: Validates concept assumptions without revealing sensitive IP.

4. Reflection + Wrap

- "Who else do you think I should talk to?"

- "Is there anyone in your team who would strongly disagree with what we discussed?"

- "If we build something that actually improves this, would you want to test it?"

Purpose: Expands your discovery network. Sets future follow-up paths.

Part 4: Live Iteration Space

Use this space to reflect after the call.

What surprised us?

(e.g., User already had a workaround. Cited a cost barrier we hadn't considered.)

What needs to change in the script?

(e.g., Add a question about how patients interact with follow-up instructions.)

What assumption got challenged?

(e.g., We assumed the surgeon felt ownership over follow-up care. They don't.) **Example: Live Script Iteration Timeline**

Startup Context: You're developing a device that removes plaque directly (not compresses it), aimed at reducing restenosis post-angioplasty. You're early in customer discovery and testing alignment with real-world clinical pain points.

Interview 1: Interventional Cardiologist

Key Feedback:

"If I am understanding your design correctly, this may very well be a substantial improvement for treating plaque build-up at bifurcation points; a common issue I see in practice is that stents simply cannot be placed at many bifurcation points."

What We Learned:
This insight introduced a **use case we had not previously considered**. The device might address a need we weren't targeting: non-stentable bifurcations. This is the kind of unexpected alignment that only emerges through open-ended discovery questions.

Script Adjustments:

- Add: "How often do you encounter plaque at bifurcation points where stenting isn't an option?"

- Add: "What's your typical approach in those cases today?"

- Add: "Would you say this type of presentation leads to worse outcomes, delays, or higher recurrence?"

Interview 2: Surgeon

Key Feedback:

"Based on what you've described, it sounds as though the guidewire and guidewire sheath are not geometrically centralized; without centralizing those components, there's a significant chance of kinking the line and reduced maneuverability."

What We Learned:

This was a **technical risk** we hadn't accounted for. The design may compromise maneuverability (particularly in tortuous vessels) if it lacks geometric centrality. This insight flags a **killer constraint** that should be addressed before further prototyping.

Script Adjustments:

- Add: "Which design elements are most critical to preserve guidewire torque and avoid kinking?"

- Add: "Have you encountered other devices that struggled with centralization or trackability?"

- Replace vague usability prompts with:
  "What warning signs do you look for when a new catheter-based tool might be difficult to maneuver?"

Interview 3: Purchasing Manager

Key Feedback:

"This sounds as though it would certainly reduce readmissions and might even take longer to perform while still being billed to the same code. How do you incentivize admin to adopt?"

What We Learned:

We gained clarity on how **clinical value alone isn't enough**. Even with improved outcomes, devices that are **time-costly but billing-neutral** may face adoption barriers. We need to build a stronger **economic value narrative** and identify which stakeholder can champion it internally.

Script Adjustments:

- Add: "Have there been examples where longer procedures were still adopted? What made the case persuasive?"

- Add: "Which outcomes or metrics are most important when making an internal value pitch to admin?"

- Add: "What kind of data (internal or from studies) would help you push something like this through a VAC or procurement committee?"

Final Evolved Script: Week 4

This is your refined script for continued customer discovery, now sharpened by early insights:

Section 1: Context & Role Framing

1. "Can you briefly describe your role and how you interact with patients undergoing angioplasty?"

2. "In your experience, how frequently do plaque buildups occur in bifurcation regions?"

Section 2: Pain Points and Unmet Needs

3. "What's your current approach when a stent can't be placed due to vessel geometry?"

4. "Have you seen poor outcomes or workarounds in those bifurcation scenarios?"

5. "Are there tools you wish existed for those specific cases?"

Section 3: Prototype Framing (Carefully)

6. "We're working on a device focused on physically removing plaque, rather than compressing it with a stent. We originally thought its main value was preventing recurrence, but now we're wondering if it could serve harder-to-treat cases like bifurcations. Based on that premise, what's your reaction?"

7. "Would a device like that raise concerns around sheath kinking or loss of maneuverability?"

8. "What design principles do you trust most to maintain torque control during difficult navigation?"

Section 4: System-Level Fit

9. "If this tool added a few minutes to the procedure but prevented repeat interventions, how would that be received by admin or VAC?"

10. "What metrics matter most when evaluating new devices under existing CPT codes?"

11. "Are there recent tools that succeeded or failed due to economic framing?"

Section 5: Expansion and Debrief

- "Who in your organization is most influential during new device evaluation?"

- "Would you be willing to test a prototype or offer further feedback in future iterations?"

IF YOU'VE NEVER EMAILED A DOCTOR BEFORE...

A Student's Guide to Cold Outreach That Actually Works

Whether you're trying to get a physician interview for customer discovery or access to a clinic to observe workflows, your first instinct might be to apologize for your lack of credentials or hope someone feels bad enough for you to say yes.

Don't do that.

Instead, **leverage the most valuable asset you already have**: your student status.

Professors, doctors, and even busy executives have a soft spot for curious, proactive students, especially those doing more than just asking for help with homework or recommendation letters. If you're in a college town where hundreds of students are fighting for shadowing hours, **your entrepreneurial ambition instantly sets you apart**. You're not just asking to tag along. You're trying to understand, innovate, and solve real problems.

Here's how to turn that into a "yes."

Cold Email Commandments

1. **Keep it under 120 words**

- No one wants a novella. Be respectful of their time and get to the point.

2. **Use a real subject line**

    - Avoid vague titles like "Request" or "Question." Instead, write:

        - Student innovator seeking 15 minutes on X

        - Quick ask: Can I learn from your experience with [topic]?

3. **Lead with who you are, not what you want**

    - "I'm a biomedical engineering student working on a project to improve surgical workflow efficiency" is better than "I'm hoping you can meet."

4. **Show specificity, not flattery**

    - "I read your paper on [X] and was intrigued by your comment on [Y]" goes much further than "I admire your work."

5. **Make the ask easy**

    - Suggest a 15-minute call or a short in-person conversation during a convenient time block. Offer to adapt entirely to their schedule.

6. **Don't over-sell or under-sell**

    - Avoid phrases like "revolutionary idea" or "just a student." Be earnest, clear, and confident.

Example Email Template (Physician)

Subject: Student innovator hoping to learn from your workflow insights

Hi Dr. Nguyen,

My name is Sara Patel, and I'm an undergrad at [University Name] working on a clinical innovation project focused on patient handoff communication. I came across your recent article on post-op complications and noticed your emphasis on information gaps during shift changes.

I'm not a company or vendor: just a student eager to learn. I'd be incredibly grateful for 15 minutes of your time to ask about the real challenges you face in this space. I'll work entirely around your schedule.

Thank you so much,
Sara

Example Email Template (Clinic Manager)

Subject: Can I shadow for 20 minutes to learn about your process?

Hi [Name],

I'm a student at [University] researching medication reconciliation workflows for a startup incubator program. I'm trying to better understand where things fall through the cracks.

Would it be possible to shadow a staff member or chat briefly with you to see how things currently operate? I know your team is incredibly busy, so I'd stay out of the way and only observe. Even 20 minutes would help me learn.

Thanks so much for considering,
Jordan Lopez

❋ **BMTT Tip: You're the Underdog; That's Your Advantage**

Most students never hit send. They worry they're too young, inexperienced, or unworthy of someone's time. But your lack of polish is exactly what makes you **low-pressure and likable**. You're not pitching a product yet. You're asking to understand. And that's disarming in the best way.

In fact, you'll often hear:

"I said yes because no one's ever asked me that before."

Or:

"I wish more students did this."

Send the email. Then send it again, if needed, to someone else.

You're not selling snake oil. You're giving smart people a chance to help fix the problems they live with daily.

CASE REFLECTION: A GOOD IDEA FOR THE WRONG PERSON

Sami's team had it all: an elegant pill dispenser prototype for elderly patients, a slick explanatory pitch, and dozens of hours shadowing nurses at local clinics. Early feedback had been positive. Every test went smoothly.

So when they launched their first large-scale discovery campaign, they expected more of the same. Instead, they got friction.

A geriatric nurse practitioner at a Medicaid-focused facility laughed when shown the demo.

"This would be amazing… for a different population. My patients don't live alone. Their daughter, neighbor, or someone from church is usually managing meds."

A social worker echoed the same thing.

"It's not that your device isn't useful. It's that our patients wouldn't be the ones using it."

Sami was frustrated. "But this solves the refill timing and adherence problem."

One pharmacist paused. "Yes, but for *who*?"

It clicked.

They had validated the problem but misunderstood the user. The MVP assumed direct interaction from elderly patients who were rarely autonomous. The first prototype relied on large buttons and audible reminders, expecting patients to acknowledge each dose themselves. It was designed around the idea of independence that, in reality, most patients in this population did not have. The team had fallen into the classic trap: solving the right issue for the wrong stakeholder.

So they shifted. Not the core technology, just the framing. They rewrote their outreach. They asked about caregiver workflows and home health logistics. They learned that most adherence tracking happened via paper logs and sporadic phone calls. With that input, the pill dispenser was reframed: instead of reminding patients to log their own use, it now served as a passive record that caregivers and home health agencies could access to monitor compliance.

The MVP didn't die. It got reassigned. Within three weeks, they had a clearer picture: their real customer wasn't the patient. It was the home health agency trying to reduce hospital readmissions and needing proof of compliance to keep contracts.

Takeaway:
When customer discovery reveals that your user is someone you never considered, that's not failure; it's freedom. It's your chance to pivot before you spend another month building for someone who was never going to buy.

CHALLENGE: Run Your First Real Campaign

You've built something functional, tested it in context, and refined it through honest friction. That alone puts you ahead of most early-stage teams.

But now comes the most important milestone in this entire playbook: **structured, large-scale customer discovery.**

Not five interviews. Not ten.
Fifty to one hundred conversations, across stakeholder roles and decision layers. This is where your idea stops being a prototype and starts becoming a strategy.

This is where you stop guessing.

Step 1: Build Your Stakeholder Map

Identify the key categories of people who touch your problem or influence its solution.

Common groups might include:

- Physicians
- Nurses or techs
- Patients or caregivers
- Clinic managers or practice admins
- IT or procurement staff

- Payers or insurers

- Allied health professionals

Deliverable: Create a table with at least 6 categories. For each, list:

- What motivates them

- What barriers they care about

- How they influence (use, approve, buy, block)

Step 2: Design Role-Specific Discovery Tools

Create tailored Google Forms, survey links, or interview scripts for each stakeholder group. Your goal is not to collect blanket feedback. It's to discover nuanced, *role-dependent friction*.

Each version should include:

- 2–3 rapport-building questions

- 3–5 problem-framing questions

- 2–3 optional MVP reflection prompts (if applicable)

- 2 "killer" questions: What would stop this? Who would block it?

Deliverable: Draft at least 3 role-specific discovery forms or script variants.

Step 3: Launch Your Discovery Campaign

Now go live. Use a combination of:

- Cold email (see templates in this chapter)
- In-person outreach on campus or in clinics
- LinkedIn messages or warm intros
- Existing university, hospital, or alumni networks

Track your outreach. You're not sending a few feelers. You're running a campaign.

Deliverable: Start a campaign log. Include:

- Name / role / institution
- Date contacted
- Response status
- Interview date (if scheduled)
- Key themes after each interview

Bonus tip: Make a "heat map" of insights. If 5 nurses mention patient confusion, that's a hotspot. Highlight it.

Step 4: Update Your Script Weekly

Customer discovery is iterative. Don't fall in love with your questions. Adapt.

Each week:

- Meet with your team
- Identify patterns
- Replace weak questions
- Add deep dives on emergent themes

Deliverable: Keep a "script changelog." Each update should include:

- What changed
- Why it changed
- What you learned that prompted it

Step 5: Extract 5 Uncomfortable Truths

At the end of 50–100 interactions, you'll have gold.

But only if you're willing to hear it.

Write down:

- 5 things that surprised you
- 3 things that made you doubt your MVP
- 2 things that made you rethink the problem itself

Deliverable: Write a one-page summary:
"**What we thought, what we learned, what we changed.**"

Anchor This

This is your crucible moment.

This campaign is not about proving your idea is great. It's about proving whether it's *necessary*. Whether it *belongs* in the system you're trying to enter.

Most failed startups trace back to skipped discovery. You're not skipping it. You're owning it.

So go out. Get 100 answers.
Learn where your idea fits and where it doesn't.
Then come back ready to build what the world actually needs.

CHAPTER 16: REFINING WHAT YOU'VE GOT

When to iterate, reframe, or kill the idea
Integrating feedback into design without becoming reactive
Deciding what your version 1 actually needs to prove

You've done the hard work.
You've gathered early testing feedback, built a first-pass product, and run structured customer discovery with key stakeholders. Now what?

This is the phase that separates momentum from motion. It's easy to keep building, tweaking, and reacting without ever stepping back to ask the bigger questions:

- Is this still *the right problem?*
- Is our solution *converging on value or drifting into noise?*
- Does our version 1 actually need to look like this, or are we over-building?

In this chapter, we'll help you navigate the messy, critical period where input turns into action. That means **deciding what to build next, what to abandon**, and **when to rethink the problem itself**.

This is not about pleasing everyone. It's about strategic refinement: making sure that your next move is not just progress, but meaningful progress.

1. ITERATE, REFRAME, OR KILL: HOW TO KNOW THE DIFFERENCE

Iteration: When the problem is still valid, but your approach needs refinement.

Signs that iteration is appropriate:

- You received consistent stakeholder interest, but they surfaced usability or workflow concerns.

- A prototype was understood and welcomed, but didn't fully fit into existing systems or space constraints.

- The pain point still resonates, and minor tweaks would unlock major improvement.

⊗ **Example:**
A wound-care device that received praise from nurses, but was too wide to fit into bedside carts. Adjusting the form factor or casing may solve it.

Reframing: When you're solving the wrong layer of the problem, or too small a slice of it.

Reframing is required when:

- Stakeholders agree the problem is real, but don't agree that your solution addresses the core of it.

- Your solution solves a symptom, but the root cause remains.

- Discovery feedback consistently highlights a *different* pain point than the one you targeted.

⊗ **Example:**
A student team builds a fall-detection wearable for seniors, but discovery reveals the real friction is in post-fall emergency response, not detection. The core need shifts from "monitoring" to "communication and response orchestration."

Kill the Concept: When neither the problem nor the proposed solution has meaningful traction.

Some signs it's time to let go:

- No stakeholder group is excited about using, buying, or adopting your solution.

- Feedback highlights solutions already on the market that perform similarly or better.

- Discovery participants express confusion about why this even needs to exist.

This is not failure. This is *founder maturity*. Many of the most successful innovators killed multiple concepts before landing on one that mattered.

2. DON'T BE REACTIVE: INTERPRETING FEEDBACK STRATEGICALLY

Just because one surgeon said something harsh doesn't mean you need to pivot. Just because one nurse said "this is brilliant" doesn't mean you're done building.

Here's how to make sense of what you're hearing:

Pattern Recognition Over Anecdote

Single quotes are valuable when illustrative, but useless when isolated. **Look for repeated patterns**, especially across stakeholder types. If three very different people said the same thing in different ways, that's a signal.

※ **BMTT Tip:** We recommend tagging feedback into thematic buckets (e.g., usability, workflow, cost concerns, implementation hurdles). Track how often each theme is mentioned across interviews. This allows you to prioritize changes that will have the broadest downstream impact.

Hold to the Problem, Loosen the Solution

It's easy to get attached to the thing you built. But don't confuse your prototype with your purpose.

You are not here to *protect* your prototype. You are here to *serve the need*. When in doubt, return to the original pain point you committed to solving and ask:

- Does this version of the product still solve that need?

- Does it solve it in a way the user understands, values, and can implement?

If not, it's time to revisit your assumptions.

3. WHAT SHOULD VERSION 1 PROVE?

One of the biggest mistakes early teams make is thinking their **Version 1** needs to be a fully featured, polished, user-ready product.

It doesn't.

What it needs to be is **a functional hypothesis**. That means something you can put in front of stakeholders to test one or more of the following:

- **Will they use it?** (Adoption behavior)

- **Can they use it?** (Usability and workflow fit)

- **Do they get it?** (Communication and positioning)

- **Will they pay for it or push for it?** (Buying signals)

Your goal is not to "launch." It's to learn. This stage is about **de-risking the core** one layer at a time.

4. TIPS FOR ITERATING WITHOUT LOSING FOCUS

Use "Throwaway" Tools with Purpose

Low-fidelity prototypes, no-code apps, foam models, and even clickable mock-ups are perfect tools here. They keep you from overcommitting too early, while still generating insight.

Good throwaway tools can answer questions like:

- Is this interaction intuitive?

- Do users recognize the value proposition?

- Does the feature set overwhelm or empower?

※ **BMTT Tip:** Throwaway tools are most powerful when accompanied by structured observation. Ask your users to "think aloud" while interacting. What do they assume it does? What do they fear? What do they ignore?

Ask Better Follow-Up Questions

This is where your skills from the last two chapters come back into play. Don't just show the prototype and ask "what do you think?" Ask:

- What part of this surprised you?

- What part would you change first?

- How would this fit into your day?

- What would make you *not* use this?

You're not asking for praise. You're asking for *obstacles*.

5. A NOTE ON TEAM DYNAMICS

Iteration doesn't just change the product. It challenges the team. You may disagree on what feedback to prioritize or whether to pivot. Be clear about your shared criteria:

- Are we chasing feasibility, usability, or adoption first?

- What tradeoffs are acceptable for Version 1?

- How many cycles of refinement will we allow before we reevaluate the concept entirely?

Be explicit. Set internal deadlines. If nothing else, it helps keep momentum from turning into inertia.

6. WHEN IS IT TIME TO ADVANCE?

You'll know it's time to move forward toward funding, partnerships, or pilot testing when:

- You've validated that at least one user group will **use** the solution in its current form.

- You've addressed the most critical usability or workflow frictions.

- You've gathered enough feedback to confidently build a **Version 1 that works in context**.

- You've done the due diligence to understand who benefits, who pays, and who blocks adoption.

In other words, your idea doesn't just work in theory; it's **proven to be worth proving more**.

STAYING IN THE GAME: WHAT TO DO IF YOUR FIRST IDEA FAILS

It's easy to romanticize the startup that gets it right on the first try. But the truth is, most ventures either pivot radically or quietly dissolve before they're ever seen. If you're staring down the possibility that your idea won't survive, or if you've already made the hard call to shelve it, know this:

You didn't fail. You just got your first real taste of how this game is played.

Pivoting Gracefully, Not Desperately

If you've gathered genuine user insight, mapped the landscape, and still find your concept dead in the water, that's data. Not defeat.

Use it to ask:

- Is there a smaller problem inside this larger one that's worth solving?

- Did you miss the real customer segment with the most urgency or budget?

- Could the core technology serve a new market entirely?

Many of the best startups emerge as spinouts of previous failures. Keep your network, notes, and insights. You may be closer to a fundable pivot than you think.

Saving Relationships, Even if the Startup Ends

Whether you had a team of two or ten, it's your responsibility to preserve those relationships. If you're pausing the project or walking away:

- Hold a debrief. Talk about what went right and wrong.

- Make people feel seen. Thank them for what they contributed.

- Don't ghost anyone. Nothing burns bridges faster.

Even if your idea isn't moving forward, your reputation will.

Preserving Momentum Without the Burnout

The biggest emotional hit isn't usually the failure; it's the stall. The limbo. The creeping self-doubt that maybe you weren't cut out for this.

You are. But founders need multiple runs to get it right.

That's why we designed this playbook to be repeatable. Every tool here can be reused, recycled, and rebuilt on. Take a few weeks to reset. Then pick a new need. Restart your canvas. Ask smarter questions. You'll move 5x faster the second time.

From Flop to Funded: Famous Pivots That Started with Failure

Sometimes your "failure" is just the prototype of your future success. These companies started with one idea and became world-changing after the pivot.

Company	The Flop	The Pivot
Slack	An online game called *Glitch*	Internal messaging system became the product
Instagram	A check-in app called *Burbn*	Focused entirely on photo sharing
YouTube	A video dating site	Became the internet's largest video platform
Twitter	A podcasting platform called *Odeo*	Pivoted after Apple launched iTunes podcasts
Airbnb	Selling cereal boxes to fund their rent	Validated hosting strangers as a service
Shopify	An online snowboard shop	Built a storefront tool for others to use

Takeaway: You don't have to force your first idea to work. You just have to stay in the game long enough to find the one that does.

CASE REFLECTION: THE LINE BETWEEN LISTENING AND LOSING YOUR WAY

Tasha's team had spent a month collecting feedback on their prototype: a physical therapy tool designed to help stroke patients retrain grip strength at home. It looked good on paper. It even won a small student pitch contest. But every interview seemed to uncover a new problem.

One therapist said the resistance bands would be too hard for early-stage rehab.
Another worried about cleaning between patient uses.
A caregiver pointed out that patients might forget how to use it without supervision.
And one older patient, blunt as ever, said, "I wouldn't use this even if it was free."

By week five, the team had 32 notes in their discovery doc: almost all critical. And worst of all, none of it aligned. Everyone had a different suggestion, a different complaint, a different idea of what "better" looked like.

Tasha wanted to build a new version. Her teammate Diego thought they should scrap it altogether. Noor argued the problem was still worth solving; they just hadn't found the right slice of it yet.

They paused.

Instead of tweaking the prototype again, they built a feedback matrix: four buckets, dozens of interviews, one color-coded sheet of truth.

That's when they saw the pattern.

Most therapists didn't mind the design. They hated the *context*. Patients were being sent home too soon, without a clear plan. The problem wasn't the device. It was that patients didn't know *when or how often* to use it.

They hadn't built a bad tool. They'd built an unscheduled one.

The next week, they reframed their entire concept: same device, new purpose. A guided progression app paired with the tool to send daily routines from the therapist directly to the patient. No guessing. No memory needed. Just do what the screen says.

Now therapists were nodding. Now patients were saying, "I could try that." Now the feedback meant something.

They didn't kill the idea.
They didn't blindly iterate.
They reframed with intention.

Takeaway:
Listening without a lens leads to noise. But listening through a structure reveals signal. Sometimes the best way forward isn't building again; it's re-understanding what people actually needed from the start.

FINAL THOUGHT

Refining your solution is not about chasing perfection. It's about ensuring that you are building something real for real users, with real constraints.

You are not here to validate your ego. You are here to validate your impact.

The next chapter will walk you through how to build early traction from the insights you've earned and how to design pilot efforts that do more than test the product. They test the **path forward**.

CHALLENGE: Make the Call

You've prototyped, tested, and listened. You've run real discovery campaigns and faced real friction. Now comes the hard part:

Making the call.

This challenge isn't about building. It's about deciding (clearly, strategically, and without flinching) what happens next. You are no longer validating a hunch. You are now refining (or retiring) a direction based on evidence.

Your task is to determine whether you should:

1. **Iterate:** Tweak and test again

2. **Reframe:** Zoom out or sideways to hit a deeper pain point

3. **Kill:** Let it go and free yourself to chase something better

Step 1: Sort Your Feedback into Patterns

Go back through your testing notes, discovery logs, and MVP trials.

Create four buckets:

- **Usability complaints**

- **Workflow fit issues**

- **Value misunderstandings**

- **Disinterest or irrelevance**

Deliverable: Create a feedback matrix. Each row = one interview or test. Columns = the four categories above. Color-code each cell with:

- Positive

- Neutral/unclear

- Negative/friction

Count your totals. You're not hunting for anecdotes. You're measuring patterns.

Step 2: Revisit the Problem, Not the Product

Write down the original **need statement** you committed to solving.

Then answer, in one sentence:

"Does our current MVP still directly and convincingly address this need?"

If you hesitate, your product may have drifted from the problem.

Deliverable: Write a 2–3 sentence summary:

- Here's the problem we set out to solve.

- Here's what users actually seem to need.

- Here's how our solution is (or isn't) aligned.

Step 3: Define What Version 1 Must Prove

Decide what your MVP absolutely needs to prove (no more, no less).

Choose **two** from this list:

- ❑ Willingness to use
- ❑ Ability to use
- ❑ Clear understanding of purpose
- ❑ Workflow compatibility
- ❑ Perceived value
- ❑ Future payment or procurement interest

Deliverable: Write one sentence for each:

"We will consider Version 1 successful if _____."

Step 4: Choose Your Path: Iterate, Reframe, or Kill

Use your matrix, your summary, and your success criteria to choose a path.

Deliverable:

- If **Iterate**: List the 3 biggest usability or adoption issues you'll address next. Build a new prototype plan.

- If **Reframe**: Write a new need statement that reflects the deeper problem or broader opportunity users revealed. Explain how it shifts your direction.

- If **Kill**: Write a 1-paragraph post-mortem. What you learned, what you'd do differently, and one spinout idea worth exploring next.

This decision doesn't need to be perfect. It just needs to be made *intentionally*.

Step 5: Lock Your V1 Build Plan

If you're moving forward, finish with this:

- What **exactly** will your next prototype test?

- What **stakeholders** will you show it to?

- What **timeline** will you follow?

- What **success signal** tells you it's ready for pilot?

Deliverable: Write a short build brief:
"**In the next 30 days, we will build a version that proves _____, test it with [X] users, and consider it successful if [Y] happens.**"

Anchor This

This is your moment of truth.

Not because you're "done," but because you've earned the right to decide what's worth doing next.

Too many teams get lost in motion by refining forever, listening endlessly, or defaulting to optimism. But you're not guessing anymore. You're building forward *on purpose*.

Whether you iterate, reframe, or walk away, own that choice.

And if you choose to move forward, do it with a smaller scope, clearer metrics, and sharper instincts than ever before.

Part VII

Understanding
The Terrain Ahead

CHAPTER 17: INTELLECTUAL PROPERTY FOR FIRST-TIME FOUNDERS

Read this, no matter your domain. Whether you're designing a medical device, a consumer app, or a wearable that tracks hydration, intellectual property (IP) is one of the few things you can own outright and mishandle completely.

This chapter won't make you an IP attorney, but it will arm you with the strategic understanding required to avoid early missteps and protect what you're building. IP can give you leverage, scare off copycats, and serve as an asset in fundraising and acquisition. But most importantly, it can help you buy time: time to build, to iterate, and to grow your moat.

1. WHAT IP IS AND IS NOT

Intellectual property refers to creations of the mind that can be legally protected. The four main categories are:

- **Patents:** Protect inventions (devices, systems, methods) for 20 years from filing.

- **Trademarks:** Protect brand identifiers like names, logos, and slogans.

- **Copyrights:** Protect original works of authorship (text, music, software code).

- **Trade Secrets:** Protect confidential business information, like formulas or algorithms, as long as you keep them secret.

IP **is not** a magic shield. It won't automatically stop others from copying you; it simply gives you legal grounds to act when they do. And it won't turn a bad product into a good one. But it *does* give you strategic options. And in early-stage startups, options are everything.

2. PROVISIONAL VS. NON-PROVISIONAL PATENTS

One of the most common questions is whether to file a provisional application first. A well-prepared provisional can secure your priority date and buy you a year to refine your invention if it's drafted seriously.

Provisional Patent Application

- **Purpose:** Locks in your filing date and lets you use "patent pending."

- **Cost:** Government fees are $75–$300, but attorney fees often run $1 K–$4 K.

- **Formality:** The USPTO does not examine provisionals, but they must sufficiently describe all inventive features to support your later claims. An incomplete or sketchy provisional can leave gaps that a non-provisional cannot fill, losing you priority rights.

- **Best Practice:** Retain counsel (university TTO, in-house, or outside) to draft a full disclosure of your invention. Many firms will offer a discount on the subsequent non-provisional filing if they prepared your provisional, since they already have your specification and drawings on file.

Non-Provisional (Utility) Patent Application

- **Purpose:** The formal examined application that may issue as an enforceable patent.

- **Cost**: Government and attorney fees often range $5 K–$15 K or more.

- **Timeline**: Examination typically takes 18–36 months.

Why Start with a Provisional

- It secures your priority date while you raise money, build prototypes, or run pilots.

- It gives you one year to test assumptions before investing in the full-blown application.

- It lets you legitimately wear "patent pending" in pitch decks, investor updates, and pilot contracts.

When to File

- **Immediately before** any public disclosure (conference demos, journal publications, or investor pitches).

- **Once** you have a clear, working description of your core invention and any critical variations.

When not to file

- If your invention is still conceptual or the key inventive aspects are undefined: you'll only trap yourself in a weak disclosure.

- Without professional help: self-filing without adequate technical and claims drafting often leaves holes that cannot be patched in the later non-provisional.

✴ **BMTT Tip:**
A provisional is only as good as its disclosure. Treat it with the same rigor you would a full application. A small up-front investment in legal help can avoid big headaches (invalidated claims, lost priority, or holes that competitors can exploit) down the road.

2.1 BOOTSTRAPPING YOUR PROVISIONAL FILING AND PITCHING SAFELY

Filing a strong provisional can cost $1 K–$4 K in attorney fees, which is often out of reach for scrappy student teams. Fortunately, well-run pitch competitions and grants exist expressly to underwrite those early IP costs as long as you frame your submission without giving away unprotected details.

How to Fund Your Provisional via Competitions

- **Seek the right contests.** Look for university-affiliated, state, or federal programs that explicitly cover "patent readiness," "IP commercialization," or "technology development." SBIR/STTR pre-proposal grants, local innovation councils, and many business plan competitions will allow you to earmark awards for provisional fees.

- **Budget your ask.** When you apply, include a simple line item ("$1,500 for provisional patent preparation")and attach an estimate from a patent attorney or your tech-transfer office. Reviewers want to see you've costed it out.

- **Leverage matching programs.** Some incubators or accelerators offer dollar-for-dollar matches on IP spending. Even if you win only $750, you can stretch that to $1,500 by pairing it with your seed or personal funds.

Pitching Without Exposing Your Invention

You do not need to show working prototypes, CAD files, or chemical structures to win. Focus your public pitch on:

1. **Problem & market validation.** Present customer-interview outcomes ("30 clinicians cited X as a daily frustration; 80 % would pay for a fix").

2. **Business model & go-to-market.** Lay out your pricing, sales channels, and revenue projections. A clear path to customers is more compelling than any half-baked demo.

3. **Competitive landscape.** Show a simple 2×2 chart or table that contrasts your approach (non-confidentially) against incumbents or academic protocols.

4. **Milestones & team.** Emphasize your traction (pilot commitments, LOIs) and the expertise of your founders. Competitions reward credibility and execution plans as much as raw invention.

※ **BMTT Tip:** If an application form asks you to "describe your tech," use high-level functional language ("a closed-loop pressure control catheter system") rather than revealing internal algorithms, exact materials, or manufacturing steps.

By combining a competitive pitch strategy with targeted funding requests, you can secure your provisional filing (and "patent pending" status) without ever posting IP online or leaking the precise workings of your invention. The result is breathing room to refine your concept under proper counsel, while you continue proving product-market fit in public.

3. FREEDOM TO OPERATE VS. PROTECTION

Freedom to Operate (FTO) means you have the legal right to build and sell your product without infringing on someone else's patent.

Protection means you've filed your own patent and have exclusive rights (if it gets approved).

They are not the same. Many startups think, "We filed a patent; we're covered." But you can still be sued for infringing on others, even if you have your own patents.

FTO requires a **search and legal opinion** to identify patents you might infringe on. This can be done by patent attorneys or consultants, especially if you're entering a crowded field (like diagnostics or wearables).

※ **BMTT Tip:** Founders often confuse having IP with having freedom. Build both. Patent smartly, but always scan the competitive IP landscape.

4. UNIVERSITY TECH TRANSFER OFFICES: FRIEND OR FOE?

If you're a student innovator working within a university, chances are your school has a **Technology Transfer Office (TTO)** or **Office of Research and Sponsored Programs (ORSP)** that handles inventions.

Here's what matters:

- **Who Owns the IP?** Depends on funding, employment, and use of university resources.

- **You likely own your IP if:**

- You're an undergraduate.
- You developed your project on your own time.
- You didn't use university-funded labs or grants.

- **You may NOT own it if:**

  - You're a graduate student or postdoc employed on a federal grant.
  - You created the invention in a university-funded lab.
  - You were paid to work on the project (RA, fellowship, etc).

※ **BMTT Tip:** Most universities have an IP disclosure process. Don't avoid it; use it to get clarity. Early conversations can help you negotiate licensing, co-ownership, or personal rights. Avoiding it can tank future deals.

※ **BMTT Case Snapshot:** We've seen multiple teams waste 6–12 months negotiating with their university after incorrectly assuming they had full rights. Early transparency saves time.

---

5. FILING INTERNATIONALLY, OPEN-SOURCE TRAPS, AND OTHER COMMON MISTAKES

International Patents:

- A U.S. patent protects you *only* in the U.S.

- Use the **Patent Cooperation Treaty (PCT)** to reserve rights in multiple countries.

- Be strategic: filing in 10+ countries is expensive. Choose based on where you plan to sell or manufacture.

Open-Source Issues:

- If you use open-source software, understand the license. GPL and similar licenses may force you to release *your* source code.

- Don't mix open-source code into proprietary systems unless you understand the consequences.

Common Mistakes:

- Publicly disclosing your invention before filing.

- Failing to document lab notebooks and invention dates.

- Filing patents on things that aren't defensible or commercially relevant.

- Assuming a patent guarantees market success.

6. IP AS A DEFENSIVE AND OFFENSIVE STRATEGY

Your patent isn't just a bragging right. It's a lever.

- **Defensive Use:** Protect yourself against infringement, block copycats, and prevent others from patenting similar ideas.

- **Offensive Use:** License your IP to others, raise capital, or position for acquisition.

Investors care about IP when:

- You're in a competitive space with fast followers.
- Your innovation is technical and core to your value.
- You're looking to raise from institutional VCs.

If your solution is easily replicable or based on process, trademark and trade secret strategy might matter more. But if your moat is technical, IP is part of your valuation.

CASE REFLECTION: PROTECT OR REGRET

Real IP stories. Real consequences. Your next move matters.

You've seen the technical side of IP, but let's make it personal. The following snapshots are based on real-world student and early-stage founder experiences. In each, we've stripped away names and institutions but kept the lessons intact. Read through, then reflect using the questions provided.

Snapshot 1: Protected and Prepared

A diagnostics team filed a well-prepared provisional before publishing early lab data. They built up a strong experimental record, then converted to a utility patent just before their Series A. Their IP portfolio passed due diligence without issue and became a key value driver in negotiations.

Think About It:

- Have you documented your technical progress with enough detail to support future claims?

- If investors looked at your IP status today, would it help you or raise red flags?

Snapshot 2: Posted, Then Panicked

A hardware startup publicly demoed their product at a university competition and later posted the footage online. They hadn't filed a provisional. That video counted as public disclosure. Their U.S. patent was still allowed, but international rights were lost forever.

Think About It:

- Have you posted or shared anything that might count as public disclosure?

- Are you tracking what's been shown, where, and to whom?

Snapshot 3: Design Wins the Day

A wearable startup was still refining its utility claims but filed early for design patent coverage to protect the visual shape of their device. They used that IP to license their design to a fashion-tech brand, generating early revenue while continuing to develop their core tech.

Think About It:

- Are there parts of your invention (appearance, layout, branding) that can be protected earlier or more easily?

- Could design or trademark filings create early value while your technical invention matures?

Snapshot 4: "But We Came Up With It!"

A student team assumed they owned the rights to their invention. They had developed the project while being paid on a federally funded grant. Turns out the university owned the IP. It took nearly a year to sort out the licensing terms before they could incorporate their startup.

Think About It:

- Have you reviewed your school's IP policy or disclosed your invention to the tech transfer office?

- Could funding, lab space, or employment complicate your claim to ownership?

FINAL NOTE: YOU CAN'T IGNORE IP, BUT YOU DON'T HAVE TO FEAR IT

You don't need to become a patent lawyer. But you do need to understand the basics and build early habits:

- Document your invention.

- Ask early who owns what.

- File strategically, not blindly.

- Protect your downside while building your upside.

Your idea may evolve, but your IP, if done right, can anchor your impact for years to come.

CHALLENGE: Secure Your Edge

Whether your idea is hardware, software, or workflow-driven, this challenge is designed to help you think strategically about **intellectual property** before it becomes a barrier.

You don't need to become an IP expert. But you do need to take ownership of the innovation you're building, and ensure you're not accidentally giving it away or infringing on someone else's.

Step 1: Clarify What You Own

Start with your current concept or prototype. Then answer:

- Have you **documented** the original idea's development and key technical features?

- Did any part of your design rely on **university resources, funding, or employment**?

- Have you disclosed the project to your **tech transfer office** (if applicable)?

- Have you **publicly presented** or shared the idea (online, conferences, social media)?

Deliverable: Create an "IP Reality Check" one-pager that includes:

- A short summary of what's novel

- A timeline of key development milestones

- A list of any external resources, funding, or employment that might impact ownership

Step 2: Identify Your IP Type

Go through your concept and tag which forms of IP apply:

- ❏ Utility Patent (device, system, or method)

- ❏ Design Patent (visual look or feel)

- ❏ Trademark (brand name, logo, tagline)

- ❏ Copyright (written material, code, training content)

- ❏ Trade Secret (algorithm, chemical formula, internal system)

Deliverable: Build a 2-column table:

- Left column = feature or asset (e.g., catheter geometry, app interface, logo, process)

- Right column = likely IP category (e.g., utility patent, design patent, trademark)

Step 3: Evaluate Your Disclosure Risk

If you've already discussed, demoed, or pitched your concept publicly, you may have triggered disclosure timelines.

Answer:

- Have you shown your invention in public or to people outside your team?

- Was that disclosure **recorded, published, or posted** online?

- Have you filed a **provisional patent** or signed NDAs before showing it?

Deliverable: Create a disclosure log. For each public interaction:

- Who saw it?

- What was shown?

- Was it recorded or posted?

- Was it under NDA?

If any of these could count as public disclosure, flag them. You may need to move quickly on provisional filings.

Step 4: Draft Your Defensive Strategy

Even if you haven't filed anything yet, you should outline a basic IP protection plan that covers:

- When you'll file a provisional patent (and for what)

- What parts of your design you will **not** share publicly

- What early NDAs you'll use with partners or vendors

- Whether you need a freedom-to-operate (FTO) search

Deliverable: Write a 3-bullet "Defense Plan":

1. We plan to file provisional coverage on [feature] before [date/event].

2. We will use NDAs for [specific interactions].

3. We need to conduct an FTO search in [domain] before further development.

Step 5: Use IP as a Strategic Asset

This step flips the script. Stop treating IP as a legal burden and start treating it as leverage. Choose one:

- ❏ File a provisional application using USPTO resources or your university TTO.

- ❏ Draft a simple "IP Highlights" slide to include in your pitch deck.

- ❏ Identify one competitor with patents in your space and read one of them. What gaps does your idea still fill?

Deliverable: Write a 2–3 sentence "IP Positioning Statement":

"Our solution protects [X], which enables [Y], and is currently not addressed by existing patents like [Z]. This gives us defensibility in [market or function]."

Anchor This

You don't need a finished patent. But you do need to know **what you own, how to protect it, and how to talk about it.**

Too many student-led startups lose their advantage because they either filed too soon, disclosed too early, or misunderstood what IP even is.

This is your chance to protect the engine before you hit the gas.

And remember: investors don't expect you to have everything filed.

But they expect you to know what's worth filing.

Protect smart. File strategically. Build your moat before someone else does.

CHAPTER 18: CLINICAL COMMERCIALIZATION 101 (SKIPPABLE FOR NON-CLINICAL READERS)

Laying the groundwork for regulated, evidence-driven markets

Why This Chapter Exists

If you're working in a non-clinical space, feel free to skip ahead. But if your solution touches patients, providers, labs, insurers, or health systems in any direct way, this chapter is essential.

Clinical innovation comes with a unique set of rules. It's not enough for your solution to work or generate revenue. It must prove itself across five distinct domains before being broadly adopted:

- **Intellectual Property (IP)**
- **Research & Development (R&D)**
- **Clinical Validation**
- **Regulatory Pathway**
- **Reimbursement Strategy**

These are the five pillars of clinical commercialization. Miss any one, and your venture can falter, regardless of how good your idea is. This chapter won't make you an expert in each domain. But it will help you ask the right questions, avoid classic missteps, and know when to bring in expert support.

THE FIVE PILLARS OF CLINICAL COMMERCIALIZATION

1. Intellectual Property: Defending Your Innovation

Before you talk to a single investor, build a single prototype, or present at a pitch competition, lock in your IP position. This isn't just about protection. It's about power, leverage, and longevity.

Key IP Concepts for Clinical Founders:

- **Patents** cover inventions and methods of use. For medical devices and therapeutics, this is the most important category.

- **Provisional vs. Non-Provisional:** A provisional patent is a placeholder (12-month clock); a non-provisional is the full application that can be examined and issued.

- **Freedom to Operate (FTO):** Just because you've filed a patent doesn't mean you can use your invention. You must check for active patents that your work might infringe.

- **University Ownership:** If you're a student, your school may or may not own your IP depending on how it was developed. Undergraduate work done outside funded labs is often yours. Anything tied to grants, lab resources, or faculty IP may not be. Always check your university's tech transfer policies.

Student-Specific Tip:
Students often have more IP rights than they realize but less strategic clarity than they need. Don't confuse *having the rights* with *knowing what to do with them*. Talk to both your tech transfer office and an outside patent attorney early.

2. R&D Planning: Mapping the Risk

Clinical innovation doesn't just require a prototype. It demands a **proof path**. This means building out a sequence of risk-reduction milestones that validate:

- Safety
- Functionality
- Manufacturability
- User performance
- Scalability

Common R&D Milestones:

- Bench testing (mechanical, chemical, biological performance)
- Simulated-use testing
- Cadaveric or animal studies (non-GLP ⇨ Good Lab Practice)
- Human usability testing (non-clinical)
- Design freeze and verification/validation cycles

※ **BMTT Tip:**
Don't build features. Build *tests*. Every R&D cycle should answer a question that reduces a risk. Each risk you retire brings you closer to a clinical-ready product and protects you from costly pivots down the line.

3. Clinical Strategy: Building the Evidence

If you're in the clinical space, you will need human data. But not all human data is created equal.

Two Questions to Ask:

1. What is the *primary purpose* of your first clinical study?
    - Regulatory approval?
    - Reimbursement justification?
    - Market traction for sales?
2. What is the minimum viable evidence needed to achieve that?

Study Design Types:

- **Feasibility or pilot study**: Small sample, used to prove early utility or safety
- **Pivotal study**: Designed to support regulatory approval
- **Post-market study**: Conducted after clearance to support adoption or reimbursement

Key Concepts:

- **Endpoints**: What you're measuring (safety, time to outcome, etc.)
- **Site selection**: Where you run the trial can impact speed, credibility, and cost

- **IRB approval**: Institutional Review Boards are mandatory for human studies

- **Good Clinical Practice (GCP)**: You will need to follow GCP standards eventually

Student-Specific Tip:
Students often have easier access to hospitals for shadowing, early feedback, or even observational studies. But don't confuse "access" with "approval." You still need IRB clearance for any formal human-subject research.

4. Regulatory Pathway: Picking the Right Track

You don't need to be a regulatory expert, but you do need to understand your category.

For medical devices (U.S. context):

- **Class I** = Low risk (e.g., tongue depressor) ⇨ Often exempt

- **Class II** = Moderate risk (e.g., blood pressure cuff) ⇨ 510(k) clearance

- **Class III** = High risk (e.g., implantable devices) ⇨ PMA required

Other Regulated Categories:

- **Software as a Medical Device (SaMD)**

- **Combination products** (device + drug)

- **Digital therapeutics**

How to Validate Your Path:

- Search **FDA databases** for predicate devices
- Submit a **513(g)** if unsure about classification
- Talk to a **regulatory consultant** early if you're considering Class II or III

※ **BMTT Warning:**
Founders who ignore regulatory early often pay the price in wasted R&D and failed pilots. Even Class I devices may have human factors or labeling rules. Know your class before your next prototype.

---

5. Reimbursement Strategy: Who Pays, and Why

This is the most commonly ignored and most venture-killing pillar.

It doesn't matter if doctors love your product if insurers won't cover it. If you don't have a plan for how you will be reimbursed, you don't have a product, you have a project.

Basic Terms:

- **Code** = The billing reference for your procedure or device
- **Coverage** = Whether a payer will reimburse the code
- **Payment** = How much they'll pay if coverage is approved

Two Common Paths:

- **Use existing codes**: If your product fits neatly into an existing procedure

- **Apply for new codes**: Long and expensive, but often required for novel tech

※ **BMTT Tip:**
Find your **proxy product**. How is something similar billed today? Search CMS coverage databases, talk to consultants, and look up reimbursement codes used in recent clinical trials. That's your anchor point.

Bringing It All Together: Your Clinical Checklist

Pillar	What You Need Early	When to Level Up
IP	Provisional patent, FTO search	Attorney-drafted claims, global filings
R&D	Bench test plan, risk map	GLP preclinical studies, design freeze
Clinical	Draft endpoints, IRB conversation	Protocol approval, site recruitment
Regulatory	Classification estimate	Pre-sub meeting with FDA
Reimbursement	Proxy code or billing model	Cost-effectiveness data, payer engagement

CASE REFLECTION: THE CLINIC IS NOT THE MARKET

What early founders miss when they cross into clinical territory

Case 1: Great Tech, Wrong Timeline
A student team developed a wearable to monitor surgical recovery at home. Clinicians loved it. Pilots were lined up. But they hadn't filed a provisional, didn't know their device class, and assumed insurance would pay "because it helps." Investors walked away, and the project stalled mid-pilot, caught in regulatory limbo and without a reimbursement plan.

Case 2: Smart from Day One
Another team tackled stroke rehab with a gamified glove. Before they ever pitched, they'd filed a clean provisional, identified existing CPT codes from similar neuro-rehab tools, and sketched out an early pilot study. They weren't regulatory experts, but they knew the questions to ask. When they entered an accelerator, their startup moved faster than anyone else's.

These two teams built promising clinical solutions. The difference wasn't technical; it was **readiness across the five pillars**. Founders who think of clinical commercialization as something to "figure out later" often discover the cost of delay is time, trust, and traction. You don't need to solve it all now. But you *do* need to know what terrain you're walking into and what the next smart step looks like.

As you move into this chapter's challenge, use these cases as a lens:

- Are you building something that touches the clinic?

- Are you thinking like someone who will have to **get it into** the clinic?

If you want to survive in regulated, evidence-driven markets, **you need more than a working prototype.** You need a plan that accounts for risk, regulation, and reimbursement.

FINAL THOUGHT: CLINICAL ≠ COMPANY

Just because you're building something clinical doesn't mean you have to form a company. Some ideas are better suited for **licensing**, **co-development**, or **research translation** through a partner.

The job of this chapter isn't to tell you what path to take. It's to help you understand the terrain and the weight of the journey ahead.

If your idea survives this framework, you're not just working on something cool. You're working on something that could actually *get to patients*.

And that's the goal.

CHALLENGE: Walk The Five-Pillar Gauntlet

If your idea touches healthcare (patients, clinicians, labs, payers, or providers), then it doesn't matter how good your prototype is if you can't cross the next chasm.

Clinical innovation must prove itself across five brutal, regulated, and expensive domains. And if you wait to face them until after launch, you've already lost.

This challenge will guide you through a **lightweight but high-impact assessment** of all five commercialization pillars.

Step 1: Complete Your Clinical Commercialization Snapshot

Download or create a table with these five rows:

Pillar	Current Status	What You Know	What You Don't Know Yet	Next Logical Step
Intellectual Property				
R&D Planning				
Clinical Strategy				
Regulatory Pathway				
Reimbursement				

Deliverable: Fill in each row honestly based on your current progress. Use this as your strategic map going forward.

Step 2: Run a Proxy Product Audit

Choose a product that is:

- Somewhat similar in form, purpose, or setting

- Already on the market and operating in a regulated clinical space

Look it up on:

- The **FDA 510(k) database** or **ClinicalTrials.gov**

- **CMS HCPCS code lookup** or **Medicare coverage database**

- PubMed or Google Scholar for early clinical data

Deliverable: Create a 1-slide visual or 1-paragraph summary titled:

"What We Can Learn from [Product Name]"

Include:

- Their regulatory class and pathway

- Type of clinical evidence they published

- Whether they used an existing billing code or pursued a new one

- Any R&D lessons you can borrow

Step 3: Define the First Real Risk

Each pillar has a primary early-stage question you must answer:

Pillar	First Critical Risk to Retire
IP	Has someone already patented this?
R&D	Does this function safely in its most basic form?
Clinical	Does this show promise in small-scale human use?
Regulatory	Do I know what class and submission path way this is?
Reimbursement	Does a similar product have a billing code?

Deliverable: Pick one of these five risks and build a plan to retire it in the next 30 days. Document:

- The risk you chose

- Why it's the most critical to solve now

- What action you'll take (e.g., interview, desk research, consultant)

- What success would look like (1–2 sentences)

Step 4: Schedule a Pillar Review Meeting

Even if it's just you and one advisor, host a short strategy session focused on:

- Reviewing the five-pillar snapshot

- Sharing the proxy audit

- Debating your priority risk and plan

Ask the advisor:

"Based on this, would you invest your own time, money, or reputation in helping move this forward?"

Deliverable: Write down the three most valuable takeaways from this conversation. They will shape how you refine your timeline, milestones, or even your project's viability.

Anchor This

Many student teams think the pitch stage is where they'll win or lose. It's not.

The real judgment happens long before (by systems, regulations, and institutions that don't care how compelling your deck is if you didn't do your homework).

This challenge doesn't ask you to solve the whole system. It asks you to prove you understand what's coming.

If you can face the five pillars with clarity and strategy, you're no longer just building a product.

You're building something that belongs in the clinic.

CHAPTER 19: TRANSLATING CLINICAL KNOWLEDGE INTO BUSINESS STRATEGY (SKIPPABLE FOR NON-CLINICAL READERS)

Why This Chapter Matters (For Clinical Founders)

Turning a clinical solution into a startup isn't just about proving that it works. It's about building a company that can navigate the labyrinth of clinical, regulatory, and reimbursement landscapes without losing time or capital. This chapter gives clinical innovators a critical framework for how to approach this complex, multi-year journey and make strategic decisions early on that will resonate with investors, regulators, and partners.

If you're operating in software, hardware, or service innovation outside of the clinical domain, feel free to skip this chapter.

Timeline Realism: Why Clinical Takes Time

Startups in the clinical space are not quick wins. For medtech devices, 6–7 years to market is normal. For therapeutics, a 10–14-year pathway is often optimistic.

Typical Medtech Milestones:

- **Year 0-1:** Preclinical validation, basic prototype, IP filing

- **Year 1-2:** Design freeze, GLP animal studies, usability & bench testing

- **Year 3:** FDA submission (often 510(k) or De Novo)

- **Year 4-5:** Clinical trial for pivotal evidence (if needed)

- **Year 5-6:** Regulatory clearance and early market entry

- **Year 6-7+:** Scale-up, reimbursement battle, and commercialization

Typical Therapeutics Milestones:

- **Year 0-2:** Lead compound ID, formulation, pharmacokinetics

- **Year 2-4:** IND-enabling studies, toxicology, GLP testing

- **Year 4-6:** Phase I safety in humans

- **Year 6-10:** Phase II/III efficacy trials

- **Year 10+:** NDA filing, commercialization

Takeaway:

Investors, partners, and even team members need to understand that long timelines are a feature, not a bug. You must communicate why your milestones are logical and staged, not slow and vague.

NAVIGATING PRECLINICAL RESEARCH, ANIMAL MODELS, AND HUMAN FACTORS TESTING

Preclinical Research

- **Bench Testing:** Required early to demonstrate basic function and safety.

- **Simulated Use:** Especially critical for surgical or workflow-adjacent tools.

- **Material Selection:** Biocompatibility, sterilization compatibility, and cost must be considered.

GLP Animal Studies

- **Why They Matter:** Often required for FDA submission if the product affects living tissue.

- **Design Carefully:** Include endpoint data relevant to the human use case (e.g., inflammation, degradation, patency).

- **Common Mistake:** Choosing a model that doesn't simulate the clinical condition well.

Human Factors Testing

- **FDA Requirement:** Especially important for digital, interface-based, or drug-device combo products.

- **How It Works:** Observing representative users attempting to use your device in a realistic setting.

- **What It Reveals:** Error patterns, interface confusion, unexpected barriers to adoption.

✳ BMTT Tip:
Even if you're far from FDA submission, the earlier you embed preclinical strategy into your product roadmap, the more compelling your story will be to investors.

Non-Negotiables vs. Deferrable Milestones

Early-stage founders often overspend on the wrong things. Here's how to categorize your development work.

Non-Negotiables (Do These Early):

- **IP Filing:** You must protect your invention.

- **Biocompatibility Analysis** (for implantables or ingestibles)

- **Basic Regulatory Strategy:** Know if you're 510(k), De Novo, PMA, or NDA/ANDA.

- **Target Product Profile (TPP):** Describes your product's clinical and functional goals

- **Foundational Preclinical Tests:** Bench or usability tests that prove the concept

Deferrable Milestones (Don't Rush These):

- **Custom Manufacturing:** Don't scale production until the design is frozen.

- **Cost-Reduction Engineering:** Premature optimization is a waste.

- **Full-Scale Trials:** Pivotal studies should only begin after solid pilot data.

- **Marketing Spend:** You don't have anything to market yet.

※ **BMTT Tip:**
In investor meetings, it is powerful to show that you know what *not* to do yet. A roadmap full of staged de-risking and capital-efficient experiments is more fundable than a mad dash to market.

CASE REFLECTION: REAL MISTAKES FROM IGNORING THIS STUFF

1. "We'll Figure Out Reimbursement Later."

A promising surgical device secured a $2M seed round and launched in 8 hospitals. But because they never validated a CPT code or pathway to payment, the hospital systems couldn't get reimbursed. None reordered. Dead in the water.

Lesson: Reimbursement is not just a go-to-market step. It must be baked into your product and pricing strategy from the beginning.

2. "We Built a Perfect MVP, But FDA Said No."

A neurostimulation startup built a functional prototype and conducted pilot testing on patients, only to discover they needed far more extensive safety testing for a Class III PMA submission.

Lesson: Regulatory classification determines your product roadmap. Talk to the FDA early. Consider a 513(g) submission if unsure.

3. "It Worked in Pigs, But Not in People."

An implantable device showed 100% success in a porcine model. In first-in-human trials, migration and inflammatory responses caused 2/5 patients to drop out.

Lesson: Animal models are guidance, not guarantees. The better your simulated use and user testing, the more prepared you are for real-world surprises.

FINAL THOUGHTS

A strong clinical strategy is about more than ticking boxes for FDA approval. It's about showing that you know the road ahead, and that you can travel it without wasting time, money, or goodwill.

If you're building in the clinical space, this knowledge *is* your business strategy.

In the next chapters, we'll shift back into general startup operations but carry forward the same lesson: **clear-eyed planning is more powerful than raw passion**.

CHALLENGE: Build the Clinical Compass

Clinical innovation is a long game. You're not just building a product; you're building evidence, credibility, and resilience over time. This challenge will help you translate your technical and regulatory path into a strategic, staged business plan that investors and partners can believe in.

Your goal: **Draft a 2-page Clinical Strategy Compass** that proves you understand the path ahead without overcommitting or overspending.

Step 1: Timeline Reality Check

Rebuild your timeline based on actual domain milestones. Choose the closest category:

- **Medtech** (e.g., diagnostics, implants, surgical tools)

- **Therapeutics** (e.g., small molecules, biologics, drug delivery)

- **Digital health / Hybrid** (e.g., digital therapeutics, SaMD, device+software)

Deliverable: Create a realistic timeline from now to market entry, including:

- Key preclinical milestones

- Regulatory checkpoints

- Evidence generation (bench, animal, or human)

- Commercial entry point

Use year markers (Year 0, 1, 2...) and include where you are *right now*.

Step 2: Separate the Must-Do from the Wait-and-See

Use the following table to separate your next steps:

Category	Do Now (Next 6 Months)	Do Later (12–24+ Months)
IP		
Preclinical Testing		
Clinical Study Planning		
Regulatory Validation		
Reimbursement Strategy		
Engineering or Manufacturing		

Deliverable: Fill this out with your specific project in mind. Then annotate each "Do Now" with why it matters *now* (what risk it removes, what it unlocks, or who it convinces).

Step 3: Pick One Mistake to Avoid

Choose **one** of the three case study mistakes outlined in the chapter:

- Reimbursement blindness

- Regulatory misclassification

- Misleading animal data

Write a one-paragraph memo titled:

"How We'll Avoid the [X] Trap"

Deliverable: Briefly describe:

- Why this specific trap is most relevant to your startup

- How you'll preempt it in your next 3–6 months

- What success looks like

Step 4: Build Your "Clinical Credibility Slide"

Even if you're not pitching yet, you will be. And clinical founders who can't communicate strategic clarity get passed over for ones who can.

Deliverable: Draft a 1-slide visual (or 1-paragraph summary) that includes:

- Your product category and regulatory class (tentative is okay)

- Your target product profile (1–2 bullets)

- 2–3 near-term milestones with cost or partnership assumptions

- A sentence showing you know what *not* to do yet

Tip: Smart founders get funded *because* they said "we're not ready to run a pivotal trial yet." Show maturity, not momentum theater.

Anchor This

Anyone can make a prototype. Not everyone can make a plan that lives in the real world of trials, regulators, budgets, and clinical complexity.

Your job now is to prove that your science has a business plan, your roadmap has restraint, and your strategy has staying power.

If you get this right, you're not just building a product. You're building the backbone of a clinical venture that can survive long enough to matter.

CHAPTER 20: COMMON BUSINESS MODELS IN HEALTHCARE AND BEYOND (FOR ALL FOUNDERS)

Why This Chapter Matters

Regardless of what you're building, you're going to need a business model. A good one. And not just any structure that brings in cash: you need one that fits your product, your users, your timeline, and your ability to scale.

This chapter is your crash course in understanding how revenue generation, user relationships, and adoption dynamics change depending on the model you choose. We'll explore the most common business models for startups across sectors, clarify how healthcare warps otherwise clean-cut logic, and present real-world examples that illustrate how the same product category can go in radically different directions depending on strategic choices.

CORE BUSINESS MODELS (AND WHAT THEY MEAN)

1. Fee-for-Service

What it is: A transactional model. You provide a product or service, and the user pays each time they use it. **Common in:** Clinical tools, diagnostics, consulting, simple testing services. **Pros:** Simple, familiar to buyers. **Cons:** Hard to scale without significant automation or staff.

2. Software as a Service (SaaS)

What it is: A subscription model for software tools, typically with monthly or annual pricing. **Common in:** Workflow tools, analytics dashboards, patient engagement platforms. **Pros:** Recurring revenue, easier forecasting. **Cons:** Churn risk, sales cycles can be long in healthcare.

3. B2B2C (Business to Business to Consumer)

What it is: You sell to a business or institution that serves the end user. **Common in:** School health tech, employer-sponsored wellness tools, hospital software. **Pros:** Easier scale than pure B2C, single contract = access to many users. **Cons:** Dual stakeholder: must convince both buyer and end user.

4. Licensing

What it is: You develop a piece of IP, and license it to another company for use. **Common in:** Biotech, university IP, algorithms, diagnostics. **Pros:** Capital-efficient, minimal infrastructure needed. **Cons:** Requires strong IP and legal counsel. High risk of stagnation if licensee deprioritizes.

5. White-Label

What it is: You build a product or platform that others rebrand and use as their own. **Common in:** Med-adjacent platforms, D2C wellness tools, diagnostics. **Pros:** High-margin, scalable. **Cons:** No brand visibility. You're replaceable.

6. Marketplace

What it is: You connect buyers and sellers, often taking a cut from each side. **Common in:** Locum tenens staffing, equipment resale, niche health communities. **Pros:** High upside, network effects. **Cons:** Extremely hard to get traction. Chicken-and-egg problem.

7. Hardware + Subscription (Razor & Blade)

What it is: Sell a physical product, and generate recurring revenue through consumables or data platform. **Common in:** Wearables, home testing kits, smart medical devices. **Pros:** Strong lifetime value. **Cons:** Requires solid logistics and operations. Hardware is hard.

WHAT CHANGES IN HEALTHCARE

While all of the above business models exist in healthcare, the context in which they operate changes everything. Here are the most common differences:

1. **Longer Sales Cycles**

You're often selling into institutions, not individuals. That means bureaucratic purchasing cycles, legal reviews, and endless compliance checks.

2. **Multiple Stakeholders**

Even if your tool helps patients, you may need to sell it to doctors, who need to pitch it to administrators, who need buy-in from payers.

3. **Regulatory and Legal Oversight**

Your solution may need to comply with HIPAA, FDA regulations, CE marking, or IRB protocols, even if it's "just software."

4. **Reimbursement Complexity**

In general tech, if a customer wants your product, they buy it. In healthcare, they often need insurance to agree. This makes pricing, coding, and coverage part of your go-to-market strategy.

5. Outcomes Matter

In health, your product isn't just a service. It has real impact on people's bodies and lives. This adds ethical complexity, trial requirements, and high expectations for validation.

CASE COMPARISONS: SAME PRINCIPLES, DIFFERENT PLAYS

A. Clinical Device (B2B Licensing Model)

- **Product:** A subdermal implant for drug delivery.

- **Model:** Sell the product to pharmaceutical companies who include it in bundled offerings or distribute it under their own label.

- **Strategy:** Pursue FDA clearance, partner with pharma early, align development with clinical trial timelines.

B. Med-Adjacent SaaS (B2B2C Subscription)

- **Product:** A post-op recovery tracking app.

- **Model:** Sell to hospitals, who provide it to patients as part of their discharge plan.

- **Strategy:** No FDA classification; focus on HIPAA compliance and usability. Emphasize reduction in readmissions and improved patient satisfaction.

C. Non-Clinical Wellness Product (D2C)

- **Product:** A wearable that monitors hydration.

- **Model:** Direct-to-consumer sales, bundled with app subscription.

- **Strategy:** Leverage influencer partnerships, build brand loyalty, avoid regulatory entanglements.

Bonus: What Impact-First Startups Get Wrong About Revenue

Mission-driven ventures often make a critical mistake: assuming that good intent equals adoption. It doesn't.

Common Pitfalls:

- **Undervaluing your solution:** Pricing too low or trying to be "free forever."

- **Skipping business modeling:** Not understanding cost of acquisition, customer lifetime value, or margins.

- **Assuming virality:** Believing your product will spread "organically" without a sales or distribution strategy.

What to Do Instead:

- Treat sustainability as a pillar of impact.

- Build a business model that works, then use that success to scale your mission.

- Be transparent: Investors, customers, and teammates respect a startup that balances heart with hard numbers.

※ **BMTT Tip: Map Your Business Model Visually**
Before you finalize your business model, create a visual flow of how value moves through your system. Who pays? Who benefits? Who decides? If you can't draw it clearly in 60 seconds, you're not ready to pitch it.

CASE REFLECTION: MODELS THAT MATTERED

Case 1: The Pricing Hero Who Burned Out
A team building a postnatal recovery app landed a grant and quickly launched a polished MVP. Their goal was impact, not profit, so they made it free. Hospitals loved it. But when the grant ran out, so did the team's ability to support users or scale. Without a revenue plan, the mission stalled.

Case 2: The D2C Detour That Paid Off
A med-adjacent startup developed a hydration wearable meant for hospital patients. But instead of going through lengthy B2B sales cycles, they launched direct-to-consumer while slowly collecting clinical validation. That early revenue gave them time, data, and leverage to negotiate with healthcare systems later.

Case 3: The Licensing Win Nobody Saw Coming
An algorithm-driven diagnostics team assumed they'd go SaaS. But feedback from clinical partners revealed that adoption would be faster if hospitals didn't need to implement anything themselves. They pivoted to a licensing model, and a top-10 lab chain became their first customer within months.

These founders didn't succeed just because their tech worked; they succeeded because their **business model fit their environment**. The best model isn't always the one that looks clean on paper. It's the one that respects your buyer's behavior, your team's strengths, and your timeline to traction.

Before you dive into this chapter's challenge, ask yourself:

- Who is really paying, and why would they keep paying?

- Are you prioritizing clarity or clinging to idealism?

- Are you forcing a model you like, or following the one the market needs?

In healthcare and beyond, your model *is* your message.

FINAL THOUGHT

Your business model is not a box to check. It's the living framework that dictates who will fund you, who will adopt you, and whether you'll survive past year one. Choose with clarity. Iterate with data. And always, always ask: how does this model help my user win?

CHALLENGE: Build Your Business Model Blueprint

You've got a solution. You've tested it. You've mapped the pain point. Now it's time to answer the hardest question you'll face in front of funders, partners, and even your own team:

"How does this make money (reliably, ethically, and at scale)?"

This challenge walks you through defining and defending your business model in a way that accounts for your product, user ecosystem, and the complexity of real-world adoption: especially in healthcare.

Step 1: Choose Your Model(s)

Review the 7 core business models outlined in the chapter. Pick:

- **One primary model** that will drive your early revenue

- **Optional: one secondary model** you may layer in later (e.g., licensing, white-label)

Deliverable: Write 2–3 sentences justifying your model choice. Avoid generic reasons. Tie it to:

- Who pays

- Who benefits

- What regulatory or purchasing friction exists

Step 2: Draw Your Value Chain

Use simple arrows or a flowchart to show **how value moves** through your system.

At minimum, your diagram should include:

- Who pays (and how)
- Who uses (and how)
- What triggers repeat or recurring use
- What blocks or delays revenue

Deliverable: A 1-slide or 1-page sketch titled:

"How Our Business Model Works in the Real World"

If you can't explain it clearly in under 60 seconds, you haven't nailed it yet.

Step 3: Customize for Healthcare (If Applicable)

If you're in or near healthcare, show how your model adapts to its unique complexities. Answer:

- Who is the decision-maker?
- Is reimbursement part of your model? If so, how?
- What compliance, legal, or stakeholder dynamics influence your sale?

Deliverable: A 3-bullet mini-brief titled:

"What Changes in Healthcare for Us"

Clarify what parts of your model are different because you're operating in a clinical or adjacent space.

Step 4: Preempt the Mission Trap

If your solution is impact-driven (e.g., mental health, rural access, child wellness), you're at risk of falling into the "free forever" or "scale without revenue" trap.

Deliverable: Answer these 2 questions in a short paragraph:

- How will you fund sustainability after your first grant or pilot?

- What's your revenue story and how does that support your mission, not contradict it?

Tip: The best impact-first startups lead with mission but survive on margin.

Anchor This

A business model isn't just about making money. It's about making momentum. It's the difference between a smart prototype and a fundable venture.

No matter your stage, this is your moment to stop guessing, stop hand-waving, and start drawing lines between product, user, payer, and path to scale.

This isn't an MBA exercise. It's the blueprint for your survival. Build it wisely. And revise it often.

Part VIII

Pathways To Build Forward

CHAPTER 21: FROM PROTOTYPE TO PATHWAY: VENTURE OR LICENSING?

Overview

You've got a promising prototype. You've validated its potential with users. You're starting to see the path forward. Now comes one of the most defining choices of your entrepreneurial journey:

Should you build this into a company yourself or hand it off to someone else?

This chapter unpacks the three core commercialization pathways:

1. **Self-run Startup**
2. **Partner-driven Licensing**
3. **IP Handoff or Acquisition**

We'll break down how to map your best fit based on personal bandwidth, IP strength, market timing, and the kind of founder you want to be.

1. WHAT KIND OF FOUNDER ARE YOU?

Before strategy, you need introspection. Your ideal pathway is not just about your product; it's about you.

- **Do you want to be the full-time CEO, raising capital and hiring a team?**
- **Would you rather stay on as the technical architect or scientific advisor?**
- **Are you energized by building or drained by operations and logistics?**

Your answer informs more than your job title. It shapes your funding path, time commitment, equity ownership, and future roles. Many brilliant inventors crash because they default into roles they never wanted.

❋ BMTT Insight:
Don't mistake your presence at a prototype's birth for a mandate to raise it through adulthood. Some founders are builders, others are bridge architects. Both are valid.

---

2. THE THREE CORE PATHWAYS

A. Self-Run Startup

You become the CEO (or co-founder) and lead the company.

Best fit if you:

- Want long-term leadership and equity
- Have high personal bandwidth or a strong team
- Are ready to fundraise, manage, and iterate constantly

What it entails:

- Incorporation and team formation
- Ownership of product development, go-to-market, fundraising
- Grant writing, VC pitches, hiring, milestone planning

Risks:
Highest time and emotional commitment. Failure often impacts personal finances and reputation.

Rewards:
Control, long-term equity upside, and strategic influence over how your vision is realized.

B. Partner-Driven Licensing

You license your IP to a strategic partner, studio, or corporate entity.

Best fit if you:

- Want to stay technical or step back while others scale

- Have strong IP but limited bandwidth or entrepreneurial appetite

- Have already attracted interest from partners or potential acquirers

What it entails:

- Creating a clean IP package (often with help from a tech transfer office or attorney)

- Structuring a license agreement (often royalty-based with potential milestone payments)

- Negotiating co-development or consulting roles if desired

Risks:
Limited control over execution. Licensing partners may pivot, shelve, or slow-roll your tech.

Rewards:
Passive income potential, retained academic or career bandwidth, lower operational risk.

C. IP Handoff (Acquisition or Option)

You transfer the rights to your IP in full, often at a pre-commercial stage.

Best fit if you:

- Want a clean break (or fast return)
- Are entering a new project, program, or startup
- Are pre-funding and wish to avoid commercialization logistics

What it entails:

- Valuation and sale of IP (or exclusive options)
- Possibly some ongoing role (e.g., technical advisor or consultant)
- One-time payment or milestone-based acquisition

Risks:
You lose most or all future upside. Execution may not reflect your intent.

Rewards:
Simplified exit, often used to seed future ventures or fund your next steps.

3. WHEN AND HOW TO APPROACH PARTNERS

Whether licensing or selling, timing is everything. If you approach too early, you look underdeveloped. Too late, and someone else may have taken your spot.

Signs You're Ready to Approach:

- IP is filed (ideally non-provisional)
- You have early usability or functional data
- A clear unmet need and differentiated mechanism
- Target market segments and early business model options

Who to Approach:

- **Strategic partners:** Companies already operating in your space who may want to acquire or license
- **Venture studios:** Groups that launch and run companies around sourced IP
- **Incubators and accelerators:** May connect you with commercial partners or serve as one

How to Start:

- Create a short **IP dossier** or **technical brief** (problem, solution, claims, and progress)
- Avoid full public disclosure; use NDAs or provisional filings to stay protected
- Focus on **what your tech enables**, not just how it works

4. PATHWAY MAPPING TOOL

Use this quick comparison to weigh your options:

Criteria	Self-Run Startup	Licensing	IP Handoff
Time Commitment	High	Medium	Low
Risk Level	High	Medium	Low
Control	Full	Partial	None
Upside	High	Moderate	Low (fixed or milestone-based)
Bandwidth Needed	Founder/CEO-level	Technical/negotiation role	Minimal
Ideal Timing	With MVP or early traction	With solid IP and early data	Pre-commercial, filed IP

5. COMMON MISSTEPS TO AVOID

- **Assuming you *have* to be the CEO** just because it's your idea

- **Waiting too long to file IP** before engaging partners

- **Locking into a pathway without exploring others**

- **Misaligning roles and equity** between technical and commercial team members

CASE REFLECTION: SAME PROTOTYPE, DIFFERENT PATHS

Case 1: The Reluctant CEO Who Burned Out

A PhD student built a promising surgical guidance tool, filed a provisional, and won multiple pitch competitions. Encouraged by momentum, she formed a company and became CEO despite never wanting to run a team. Six months later, investor pressure, hiring stress, and daily logistics had drained her passion. She stepped down, and the company stalled while it searched for new leadership.

Case 2: The Strategic License
A biomedical engineer created a high-accuracy diagnostic patch during grad school. Rather than incorporate a startup, she partnered early with a diagnostics company, negotiated a licensing deal through her university, and stayed on as a technical advisor. The product was commercialized within two years, and she used the income to fund her next venture.

Case 3: The Clean Break with Momentum
An undergrad team developed a novel catheter design. Knowing they lacked the capacity to run a clinical-stage company, they worked with their tech transfer office to craft a tight IP package and handed it off via a pre-seed acquisition. The buyer built the infrastructure, and the team moved on: credited, compensated, and unburdened.

These founders didn't succeed *because* they picked the same path. They succeeded because they picked the right path **for them**. Startups, licensing, and exits all require different resources: **not just funding, but energy, motivation, and tolerance for risk**.

As you approach this chapter's challenge, ask yourself honestly:

- What do I want my day-to-day to look like a year from now?

- Is my current plan based on strategy or inertia and default roles?

- Am I building a company because I *should*, or because I *want* to?

FINAL WORD

Your prototype is the seed. But your pathway is the soil. Choose where and how you plant carefully.

Whether you lead the charge, license your breakthrough, or hand off your invention to more seasoned hands, the key is **intentionality**. You don't need to do everything yourself. You just need to make sure it gets done by the right people, at the right time.

In the next chapter, we'll lay out the startup building blocks if you do choose to go the founder-led route (equity, co-founders, and early legal architecture that won't break under pressure).

Let's build smart.

CHALLENGE: Define Your Commercialization Path

You've made it past ideation, validation, and prototyping. Now comes the choice that determines not just your startup's direction, but your role in it.

This challenge helps you map your best-fit commercialization path and take the first real steps toward it.

Step 1: Choose Your Pathway

Pick one (for now):

- ❏ Self-Run Startup

- ❏ Partner-Driven Licensing

- ❏ IP Handoff or Acquisition

Deliverable: Write a 3–5 sentence statement explaining:

- Why this path fits your personal bandwidth, motivation, and skill set

- What your ideal role would be (e.g., CEO, technical founder, advisor)

- What your long-term vision is for the idea

Be honest. This isn't about what sounds impressive. It's about what's sustainable for you.

Step 2: Score Your Readiness

Use this mini self-assessment to score your current readiness on a scale from 0 to 3 (0 = not at all, 3 = fully ready):

Criteria	Score (0–3)
You've filed a provisional or non-provisional patent	
You've validated usability or functional performance	
You've outlined a business model or partner profile	
You've mapped market need and key differentiators	
You can articulate what your tech enables (not just how it works)	

Deliverable: Sum your total score out of 15. Use it as a gut check for whether you're ready to approach partners, raise funds, or seek acquisition, or if more internal work is needed first.

Step 3: Create a Go-Forward Brief

No matter your pathway, you'll need a short, clear artifact to move forward.

Choose one of the following and draft it:

- If you're pursuing **a startup** ⇨ create a 1-page founder vision brief

    o Include: the need, what you're solving, your intended role, and your 3-month next steps

- If you're pursuing **licensing** ⇨ draft a 1-page IP overview

    o Include: the problem, solution, IP status, supporting data, and ideal licensee profile

- If you're pursuing **a handoff** ⇨ outline a 1-page exit snapshot

    o Include: the tech description, current progress, value proposition, and proposed sale/option structure

Deliverable: Title the document based on your path. Think of it as your first outbound asset (something you could show an investor, partner, or studio tomorrow).

Step 4: List Your First 3 Conversations

You cannot do this alone. List three people you will contact to help you pressure-test or pursue this pathway:

- A peer, mentor, or professor who understands your space

- A legal, commercialization, or tech transfer expert (if applicable)

- A potential partner, advisor, or early-stage funder

Deliverable: Write their names and what you want to ask each person. Include how you'll reach out and when.

Anchor This

A great prototype means nothing without a pathway forward. Whether you decide to lead, license, or let go, this moment is not about ego; it's about execution.

The best founders aren't the ones who do everything. They're the ones who know what role to play, and when to hand the baton.

Choose your path. Map your next steps. And start moving.

The world doesn't need perfect founders. It needs real ones who ship.

CHAPTER 22: STARTUP BASIC FOR BUILDERS

What Every First-Time Founder Needs to Set Up Right... Before It Gets Expensive

Overview

Ideas are fragile. But the wrong legal setup is lethal.

Many first-time founders put off decisions about incorporation, equity, and intellectual property: either out of confusion or fear of commitment. The problem? Every step you take without proper infrastructure increases the risk of misalignment, legal exposure, or worse: killing your company before it gets started.

This chapter is your high-speed primer on the startup infrastructure that actually matters. Whether you're building a medtech device, a SaaS tool, or a consumer product, these decisions will shape your venture's future.

1. CHOOSING THE RIGHT LEGAL ENTITY

LLC, S-Corp, or C-Corp: What's the Difference?

Entity Type	Pros	Cons	When It Works
LLC (Limited Liability Company)	Flexible, fewer formalities, pass-through taxation	Not VC-friendly, equity complications	Solo builders, services businesses, or small licensing plays
S-Corp	Tax benefits, pass-through income	Restrictions on ownership (US citizens only, one class of stock)	Small domestic startups not planning to raise institutional capital

Entity Type	Pros	Cons	When It Works
C-Corp (usually Delaware)	Preferred by investors, multiple stock classes, scalable	Double taxation, more complex setup	Best for VC-backed or scalable product startups

TL;DR: If you plan to raise venture capital or issue stock to employees, **form a Delaware C-Corp.** If you're building a bootstrapped or lifestyle business, an LLC or S-Corp may work better, but reassess before you take funding.

2. FOUNDER EQUITY AND CAP TABLE HYGIENE

Equity feels abstract until it becomes a fight. Bad early decisions about ownership are the most common cause of founder conflict and startup death.

Common Mistakes to Avoid:

- Splitting equity 50/50 "just to be fair"
- Granting shares without vesting or contribution timelines
- Ignoring future hires or equity for advisors

Best Practices:

- **Use a vesting schedule** (typically 4 years with a 1-year cliff)
- **Document roles and responsibilities** before issuing equity
- **Create an equity reserve** (10–15%) for early team members and advisors

- **Track everything** with a proper cap table (use tools like Carta, Pulley, or Excel + templates)

❋ **BMTT Tip:** Founder equity should reflect a combination of contribution (past and expected), risk taken, and commitment. Equity is not just a thank-you card; it's a long-term contract. For a deeper dive into the complexities of founder equity decisions, see *The Founder's Dilemmas* by Noam Wasserman, which explores common pitfalls and trade-offs around splitting ownership, hiring, and control.

---

3. IP ASSIGNMENT AND UNIVERSITY TTOS

One of the biggest risks in academic-born ventures? **Not owning your own invention.**

If You're a Student:

- At many schools, **you own what you invent**, unless:
    - You were paid by a research grant or used significant university resources
    - You were on a state or federally funded project
- But ownership can still get murky; **clarify early with your Tech Transfer Office (TTO)**

If You're Faculty or Staff:

- IP created "in the course of employment" is typically owned by the university
- TTOs can license this IP to you, but they often take equity or royalties

Key Concepts:

- **IP Assignment Agreement**: Ensures IP is transferred from individuals to the company (this is essential before raising capital)

- **Employee IP Agreements**: All co-founders and hires should sign these to avoid future claims

- **Bayh-Dole Act**: Gives universities rights to federally funded inventions but allows for exclusive licensing to companies

※ **BMTT Insight:** Most founders delay this conversation until investors ask for clean IP rights, and by then, it's too late to negotiate. Handle it in the first 60 days, not the last 6 weeks before funding.

4. FOUNDER VESTING AND LONG-TERM COMMITMENTS

You've likely seen the horror stories:

- A co-founder leaves six months in but keeps 30% equity

- A falling out turns into a lawsuit over who owns what

- An investor backs out because IP isn't assigned or equity isn't vested

To avoid this:

Use a Standard Founder Vesting Agreement:

- 4-year vesting, 1-year cliff is the norm

- Optional: milestone-based vesting for part-time contributors

Discuss Exit and Commitment Scenarios:

- What happens if someone wants out?

- What if one founder is full-time and the other is part-time?

- What if one wants to license out the IP and the other wants to go all-in?

BMTT Framework:
Have an explicit conversation around:

- Roles and time commitments

- Equity split logic

- Decision-making authority

- Definitions of success/failure

- What happens if timelines shift or one founder's life changes

Write it down. Call it a "Founder Alignment Memo." Revisit it every 6 months.

5. LEGAL INFRASTRUCTURE AND LONG-TERM DEFENSIBILITY

This playbook was born from student-led ventures. But every investor, accelerator, and partner you approach will want to know the same thing:

Is this startup built to last or built to break?

Here's what you'll be expected to show:

- Your **legal entity** is clean and registered correctly

- Your **cap table** is trackable and makes sense

- Your **IP is properly assigned** to the company

- Your **co-founder agreements** include vesting, equity rationale, and exit clauses

- You're **ready to issue equity** to advisors, hires, or future investors

If you can check those boxes, you've built an investable foundation. If you can't, fix it before pitching.

CASE REFLECTION: BUILT TO WIN, OR BUILT TO BREAK?

Case 1: The 50/50 Fallout
Two friends built a diagnostics startup together and split equity 50/50 on day one "to be fair." A year later, one had gone full-time while the other became sporadically involved. The full-time founder tried to raise funding, but investors balked: why was half the company owned by someone barely present? Tension turned to legal battles. The company never raised.

Case 2: IP Assumed, Not Assigned
A student founder developed a promising wearable in a university lab and pitched it at several competitions. When an accelerator offered funding, the diligence revealed the university might own the core IP. The offer was pulled. It took 10 months and legal help to untangle the ownership, and momentum never recovered.

Case 3: The Invisible Cliff
A three-person team formed a Delaware C-Corp and built a beautiful prototype. But they hadn't put themselves on vesting schedules. When one founder left abruptly, they walked away with 30% equity: blocking future hires, souring potential investors, and forcing a messy restructuring that nearly tanked the cap table.

These stories aren't about dramatic betrayal or bad intentions. They're about **early sloppiness turning into late-stage dealbreakers**. Legal basics like entity type, IP ownership, and equity structure might feel secondary while you're prototyping, but they quietly define how fundable, protectable, and durable your startup actually is.

As you enter this chapter's challenge, ask yourself:

- Could you hand your current setup to an investor and be taken seriously?

- Have you made decisions or just left defaults in place and hoped for the best?

- If someone leaves your team tomorrow, are you protected?

You're not just building something exciting.
You're building something others might want to fund, join, or acquire.
Make sure it's something they can trust.

FINAL WORD

No one becomes a founder to do paperwork. But early mistakes with legal structure, equity, and IP don't just slow you down; they can kill your company outright.

You don't need a law degree. But you do need structure.
Build it now.

CHALLENGE: Build Your Startup's Legal Backbone

This chapter wasn't glamorous, but it was critical. If you've made it this far, you're no longer "just exploring": you're building. And what you build now will either attract investors and collaborators or repel them.

Let's make sure your foundation is real.

Step 1: Choose Your Entity (or Validate the One You Have)

Pick the entity structure that aligns with your goals:

- ❑ Delaware C-Corp (if raising VC or issuing stock)
- ❑ LLC (if bootstrapping or licensing)
- ❑ S-Corp (rare, but viable for small, domestic teams)

Deliverable: Write a 2–3 sentence justification for your choice (or your current entity). If you're not sure yet, identify the top two questions you need answered before deciding.

Step 2: Draft Your Founder Equity Plan

Use this 4-box table to outline your early equity logic.

Founder Name	Role/ Contribution	Time Commitment	Proposed Equity %

Deliverable: Fill in the table and reflect:

- Does this reflect actual risk, labor, and commitment?

- Have you included vesting (4-year, 1-year cliff is default)?

- Did you reserve 10–15% for future team members or advisors?

If not, revise.

Step 3: Check Your IP Position

Choose one:

❏ I'm a student and built the project outside of funded university work

❏ I used lab time, funding, or worked on a PI-led project

❏ I'm faculty or staff and may be subject to university IP policy

Deliverable: Based on your situation, write down:

- Whether you currently own the IP

- Whether the university might claim ownership

- One next step you'll take (e.g., schedule TTO meeting, start disclosure, draft assignment agreement)

Step 4: Draft Your "Founder Alignment Memo"

Write a short internal doc (250–400 words) that includes:

- The long-term vision

- Roles and responsibilities

- How equity was divided and why

- What happens if someone leaves or scales back

- How decisions will be made and revisited

Deliverable: This memo is not legal paperwork, but it may prevent your biggest legal disaster later. Create it now and schedule a check-in to revisit it in 6 months.

Step 5: Create a Legal Infrastructure To-Do List

Check off what you've already done, and circle what needs action:

- ❏ Legal entity formed and in good standing

- ❑ Cap table started and trackable (Carta, Pulley, or spreadsheet)

- ❑ IP assignment agreement signed by all founders

- ❑ Founders on vesting schedules

- ❑ Equity reserved for hires and advisors

- ❑ University IP questions clarified (if applicable)

- ❑ Ready to issue equity to future team or raise capital

Deliverable: Add deadlines next to any incomplete items. Don't let this list gather dust.

Anchor This

You don't need to have everything perfect right now. But you do need to show that you're not building on sand.

Funders, collaborators, and advisors don't expect you to be a lawyer, but they expect you to be prepared.

A clean setup today is your signal to the world that you're serious tomorrow. Lock it down. Then go build something worth protecting.

CHAPTER 23: STRATEGIC PLANNING AND RISK READINESS

Plan Like You're Serious, Because You Are

In the early stages of a startup, planning is not about predicting the future. It's about preparing for what might go wrong, clarifying what success actually looks like, and setting intelligent boundaries so you don't chase a good idea off a cliff. This chapter isn't about building the perfect Gantt chart. It's about defining your kill-zones, structuring your first 12 months around proof, not press, and embedding enough risk awareness into your strategy that you always know where you stand.

1. SETTING "KILL ZONES": KNOW WHEN TO WALK AWAY

The hardest decision in a startup isn't pivoting; it's quitting. Most founders only consider that once they've burned through cash, morale, and runway. **Kill zones** are pre-defined boundaries (quantitative, time-bound conditions) under which you agree to pause, pivot, or shut down.

Examples of Kill Zones

- "If we can't enroll 10 pilot users by March, we stop."

- "If our customer interviews don't yield at least five verbal commitments by Q2, we reassess."

- "If 80% of our target clinicians say they won't adopt our workflow, we kill this line of development."

※ **BMTT Tip:** Kill zones are not admissions of failure. They are guardrails against sunk-cost bias. Set them with your team and advisors before momentum blinds you to emerging warning signs.

2. YOUR FIRST 12 MONTHS: MILESTONE-BASED PLANNING

Don't plan your first year like a product manager. Plan it like a founder with proof to show. Rather than arbitrary sprints, anchor your roadmap in **validation milestones** that, once achieved, materially increase your venture's credibility.

Milestones That Matter

- **Prototype Complete:** A minimally functional version, even if rough.

- **Early User Commitments:** 5–10 letters of intent or pilot agreements.

- **Regulatory Snapshot:** A formal classification opinion (e.g., 513(g) request filed).

- **Non-Dilutive Win:** Secured SBIR/STTR or equivalent grant.

- **Institutional Interest:** Letter or meeting request from a strategic partner.

- **Revenue Signal:** First paying customer or pilot contract.

Each milestone must answer:

"If this were true, would our startup be significantly more fundable, buildable, or credible?"

Then map for each:

- **What** you must deliver to reach it

- **Who** must validate it

- **What evidence** you must capture

3. BUILD TOWARD TRACTION, NOT PRESS

Early founders often chase the wrong scoreboard like news articles, awards, LinkedIn shout-outs. Those can feel good but seldom translate into tangible progress. **Traction** is the foundation of your next raise, not the byproduct.

True Traction Signals

- **End-user Commitments:** Letters of support or signed pilot agreements.

- **Grant Funding:** Non-dilutive capital matched by your own investment of time.

- **Pilot Launch:** IRB approval or operational study underway.

- **Design Iterations:** Real-world feedback leading to measurable improvements.

- **Partner Pilots:** A hospital, clinic, or distributor formally engaged in testing.

※ BMTT Tip: If a milestone doesn't advance your funding or de-risk your technology, reconsider its priority.

4. EMBED RISK AWARENESS, NOT JUST RISK LISTS

You've learned to classify risks by likelihood and impact. Now, weave that awareness into every decision. Rather than a full matrix here, identify **your top three "must-nail" risks** (the single biggest regulatory hurdle, the core technical unknown, and the most likely team or IP challenge).

For each top risk, ask:

- **Why** is this high priority?
- **How** will we test or mitigate it in the next 30 days?
- **Who** owns that work and by when?

This lean risk-checklist ensures you never ignore the hazards that could kill your project, yet stay nimble enough to act.

5. DECIDE WHEN TO SPEND AND WHEN TO WAIT

Moving fast doesn't mean moving recklessly. Capital efficiency is a superpower in pre-seed and seed stages.

Spend When

- You've validated the underlying need and your delivery path.
- The expense will directly accelerate a critical milestone.
- You must protect IP or regulatory positioning that's time-sensitive.

Wait When

- You're still uncertain of user behavior or core use cases.
- A lower-cost simulation or prototype can answer the same question.
- The next decision is irreversible (hiring, manufacturing tooling, formal partnerships).

Ask Yourself: "Will this spend reduce uncertainty, or just improve optics?"

CASE REFLECTION: THE STARTUP THAT WAS "ALMOST THERE" (TOO MANY TIMES)

Jules co-founded a digital health startup during her final year of undergrad. Her team had a promising idea: a symptom-tracking tool for patients recovering from orthopedic surgery. It wasn't revolutionary, but it had legs: especially after two surgeons and a physical therapist agreed to help test it.

By month three, they had:

- A no-code prototype

- IRB paperwork in motion

- A pitch deck that looked decent enough for competitions

But every few weeks, they shifted focus.

"We just need to get into this accelerator."
"Actually, let's build a version for geriatric care instead."
"Wait… one of our mentors thinks it should be an AI tool."

They weren't spinning their wheels. They were always working. But when Jules' mentor finally asked to see their 6-month roadmap, the team froze.

"We've got options," they said.
"But what are you actually proving?" he asked.
Silence.

The Turning Point
That conversation forced them to write down:

- What milestone would make this startup fundable

- What user behavior they *needed* to confirm

- And what deadline would make them walk away

They landed on this kill zone:
"If we don't secure three IRB-approved pilots or equivalent user studies by September, we pause the venture."

Suddenly, their to-do list changed. They spent less time polishing and more time calling clinics. One fell through. One delayed. One said yes.

They didn't "make it" that fall. But they moved from motion to traction.

The Lesson
Planning isn't about having a perfect map. It's about knowing what matters enough to anchor you and what risks will sink you if you ignore them. Jules' team didn't fail because they didn't work hard. They almost failed because they didn't decide what counted as progress.

You don't need certainty. But you do need boundaries.
Kill zones, validation milestones, and real risk ownership won't slow your startup down. They'll keep it alive.

FINAL WORD

Founders who plan for traction and manage risk win. This chapter isn't about being overly cautious; it's about being deliberate. A good founder takes risks. A great one measures them, mitigates them, and walks away from those that don't make sense.

Next up: In the next chapter, we'll dive into the science of risk management by looking at quantitative scoring, failure-mode analysis, and living risk-registers that turn uncertainty into actionable roadmaps. Because diligence isn't just something you do; it's a discipline you master.

CHALLENGE: Turning Planning Into Proof

This chapter has shown you how to define when to stop, what to prove first, how to recognize real traction, and how to weave risk awareness into every decision. Use the steps below to translate strategy into action and ensure you're building with purpose, not just hope.

Step 1: Write Your Kill Zone

Goal: Define the single most critical boundary that, if unmet in the next 6–12 months, means you must pause, pivot, or shut down.

- **Measurable:** A clear metric (e.g., pilot users, customer commitments).

- **Time-bound:** A firm deadline (e.g., by June 2026).

- **Critical:** Directly tied to fundability, adoption, or technical feasibility.

Deliverable: One-sentence kill zone (e.g., "If we haven't secured at least five signed pilot agreements by May 2026, we pause this project.")

Step 2: Build Your 12-Month Milestone Map

Goal: Anchor your first year in 3–5 validation milestones that materially increase your venture's credibility.

Milestone	What to Deliver	Who Must Validate It	Deadline
e.g. 5 Letters of Intent	Pilot agreement template	Hospital procurement lead	August 2025
e.g. Functional Prototype	Basic working demo	Technical advisor	October 2025
e.g. SBIR/STTR Submission Filed	Grant application	Grant officer or mentor	December 2025

Deliverable: Complete the table above for your 3–5 milestones, then choose the highest-priority milestone and outline your first steps for this week.

Step 3: Identify Your Top Three Risks

Goal: Choose the three risks that pose the greatest threat to your venture right now (one regulatory, one technical, one team/IP). For each risk, answer:

1. Why is this a "must-nail" risk?

2. How will you test or mitigate it in the next 30 days?

3. Who is responsible and by what date?

Deliverable: A one-paragraph "risk brief" for each of your top three risks.

Step 4: Commit to a Smart Spend vs. Wait Plan

Goal: Decide where to invest in the next 90 days (only where it directly accelerates a critical milestone or de-risks your top risks) and identify one area where you should hold off.

- **Spend When:** Validation or risk mitigation depends on it.

- **Wait When:** A lower-cost alternative exists or you lack clarity.

Deliverable:

- **Spend:** [Activity] — 1-sentence rationale

- **Wait:** [Activity] — 1-sentence rationale

Share these with a peer or advisor to keep yourself accountable.

Anchor This:

A deliberate plan built on kill zones, proof-focused milestones, real-world traction, and lean risk awareness is your foundation for disciplined execution. Complete these steps now and you'll enter the next chapter ready to master the next level of risk-reduction science.

CHAPTER 24: THE SCIENCE OF RISK MANAGEMENT

All the diligence you have done so far (need validation, landscape mapping, milestone planning) has been one long exercise in reducing uncertainty. Now it is time to treat risk reduction as its own discipline. In this chapter you will learn how to turn qualitative instincts into quantitative scorecards, how to apply failure-mode analysis, and how to build a living risk register that guides every decision.

1. WHY RISK MANAGEMENT IS ITS OWN SCIENCE

Every startup encounters unknowns. The difference between reactive founders and disciplined founders is that the latter treat uncertainty as data. You have already identified kill zones and woven risk awareness into your 12-month roadmap. Now you will learn methods to:

- Quantify risk likelihood and severity

- Prioritize the hazards that can kill your project or slow it down

- Track and audit your mitigation efforts over time

By mastering these tools you will move from "hoping we don't hit this problem" to "knowing when and how we will address it."

2. RISK REDUCTION TACTICS ACROSS SETTINGS

Before we dive into scoring, let us survey tactics you can borrow from medical device, aviation, and software industries. Each approach shares one core principle: move uncertainty from unknown to known, and from known to solved.

2.1 Medical Device Best Practices

- **Design Controls**
  Document user needs, design inputs, and verification tests in traceable documents.

- **Pre-Submission Meetings**
  Engage regulators early to clarify classification, testing requirements, and clinical endpoints.

- **Human Factors Studies**
  Observe real clinicians using prototypes in simulated environments to uncover hidden hazards.

2.2 Software and Technology Tactics

- **Fault Tree Analysis**
  Map out logical pathways from root cause failures to system-level incidents.

- **Continuous Integration Testing**
  Automate regression tests so new code changes do not re-introduce known bugs.

- **Issue Tracking**
  Use a ticketing system for every bug, feature request, or compliance gap, tagged by severity.

2.3 Simple Startup-Scale Approaches

- **Pre-Mortem Workshops**
  Gather your team to imagine why the venture failed, then reverse-engineer safeguards.

- **Rapid Prototyping of Failure Points**
  Build a minimal version of your riskiest component (e.g., a 3D-printed catheter head) and stress test it.

- **External Audits**
  Bring in a mentor, consultant, or advisory board member to critique your top three risks every quarter.

3. QUANTITATIVE SCORING: PROBABILITY AND SEVERITY ON FIVE-POINT SCALES

A simple but powerful method is to rate each risk on two axes: probability of occurrence (PO) and severity of harm (S), each on a five-point scale:

Score	Probability of Occurrence (PO)	Severity of Harm (S)
1	Rare – Unlikely to occur in project lifetime	Negligible – Minor inconvenience only
2	Unlikely – Could occur under unusual conditions	Minor – Non-critical but requires fix
3	Possible – Known to occur in similar projects	Moderate – Impacts schedule or cost
4	Likely – Occurs regularly without controls	Major – Jeopardizes core functionality
5	Almost certain – Virtually guaranteed	Catastrophic – Safety, legal or exit risk

Multiply PO × S to yield a **Risk Priority Number** (RPN) between 1 and 25. Use this RPN to classify each risk:

- **1–6**: Acceptable

- **7–12**: Acceptable with modifications

- **13–25**: Not acceptable until mitigated

This matrix turns abstract fears into concrete scores you can compare and track.

4. FAILURE-MODE EFFECTS ANALYSIS (FMEA) TABLE

To organize your risks, build an FMEA table with these columns:

Risk ID	Hazard	Reasonably Foreseeable Sequence	Hazardous Situation	Potential Harm	PO	S	RPN	Action Required
1	Catheter over-inflation	Pressure control failure	Device head expands beyond vessel	Vessel rupture, hemorrhage, patient injury	2	5	10	Redesign pressure relief valve
2	Software crash in monitoring app	Memory leak after 10,000 patient entries	App freezes during upload	Data loss, missed infection alerts	3	4	12	Implement automated memory tests
3	IP assignment dispute	University claims ownership of patent	Funding delays during legal review	Project halt, investor pull-out	2	4	8	Confirm assignment agreements

Populate one row per critical hazard. Focus first on RPNs above 12 (not acceptable) and 7–12 (acceptable with modifications).

5. EXAMPLE: CATHETERIZED PLAQUE REMOVAL DEVICE

Let us walk through Risk ID 1 in detail:

- **Hazard:** Over-inflation of the catheter head

- **Foreseeable Sequence:** Failure of onboard pressure sensor or algorithm miscalculation

- **Hazardous Situation:** Device head expands beyond the vessel's safe diameter

- **Potential Harm:** Vessel wall stress leading to plastic deformation or rupture

- **Probability (PO = 2):** Unlikely on first prototypes but known risk without sensor redundancy

- **Severity (S = 5):** Catastrophic – patient safety and regulatory non-compliance

- **RPN = 10:** Acceptable with modifications

- **Action:** Add a mechanical pressure relief valve and dual-sensor cross-check within 30 days

6. BUILDING A LIVING RISK REGISTER

Your FMEA should live in a shared document or project management tool. To keep it alive:

- **Monthly Reviews:** Update PO, S, and RPN after each prototype test or regulatory discussion.

- **Change Log:** Record when you close a risk or adjust mitigation tactics.

- **Alerts:** Flag any RPN that rises above 12 due to new information.

- **Ownership:** Assign each risk to a team member with deadlines for mitigation tasks.

7. MOVING FORWARD: FROM SCORECARDS TO CULTURE

Quantitative risk management is not an exercise you check off once. It is the engine of disciplined execution. By scoring every hazard, tracking mitigation progress, and embedding this practice into your weekly rhythms, you ensure your startup makes data-driven decisions, not guesswork.

When you finish this chapter you will have:

- A prioritized FMEA table with real RPN scores

- A clear set of mitigation actions for your highest-priority risks

- A template for a living risk register that updates with every new insight

Mastering risk management is not about avoiding all danger; it is about knowing exactly how much you can tolerate, and when to act decisively.

CASE REFLECTION: THE RISK THEY COULDN'T AFFORD TO IGNORE

Max and Priya had been working on their medtech startup for months: a minimally invasive device to remove clots in small-vessel strokes. Their prototype was elegant. Their pitch deck had won two competitions. A neurosurgeon at a regional hospital even agreed to be a clinical advisor.

But behind the momentum was a truth they were starting to avoid: they didn't have a risk plan.
"We'll address that once we get into the accelerator," Max kept saying.
"We know the main issues," Priya would add. "They're just not top priority yet."

That changed the day they got a call from a grant reviewer who'd seen their application.

"I like what you're doing," she said. "But you flagged zero regulatory or mechanical risks. That's not confidence; it's blindness."

They realized they had no formal way to rank the dangers they knew were looming:

- The balloon head's overexpansion during deployment

- Inconsistent results in smaller vessels during early testing

- Uncertainty about how the FDA would classify their device

They'd been focused on polishing the story, yet the real story was where things could go wrong.

The Breakthrough
Their advisor insisted they sit down for what he called a "pre-mortem."
"Tell me how this startup fails," he said.
They sketched out three failure modes: a prototype flaw, a failed bench test, and an unclear regulatory path that spooked investors.

Then they built their first risk matrix:

- **Probability of failure in bench testing: 3**

- **Severity if it fails in vivo: 5**

- **RPN: 15 — not acceptable**

They added a mitigation: run exaggerated stress tests on the balloon head and design a mechanical stop to prevent overinflation. They assigned it to Max with a 3-week deadline.

The Lesson
They didn't suddenly have all the answers. But now, risk wasn't a vague feeling; it was a list, a table, a calendar alert. Something they could manage.

CHALLENGE: Turn Uncertainty Into Actionable Data

This challenge will help you apply the methods in this chapter to your own startup. By the end, you'll have a ranked FMEA table, mitigation actions for your top risks, and a stubbed "living" risk register you can update as you learn.

Step 1: Identify Your Top Three Hazards

List the three risks you believe pose the greatest threat to your venture right now (technical, regulatory, clinical, IP, etc.).

- Deliverable: A table with columns **Risk ID, Hazard, Reasonably Foreseeable Sequence, Hazardous Situation,** and **Potential Harm**. Fill in one row per hazard.

Risk ID	Hazard	Reasonably Foreseeable Sequence	Hazardous Situation	Potential Harm
1				
2				
3				

Step 2: Score Each Hazard

For each of your three hazards, assign:

- Probability of Occurrence (1–5)

- Severity of Harm (1–5)
  Then compute RPN = PO × Severity.

- Deliverable: Extend your table with **PO, Severity** and **RPN** columns.

Risk ID	...	PO	Severity	RPN
1				
2				
3				

Step 3: Classify & Plan Mitigations

Using your RPN scores, classify each risk as:

- 1–6 ("Acceptable")
- 7–12 ("Acceptable with modifications")
- 13–25 ("Not acceptable until mitigated")

For each risk, write a one-sentence **mitigation action** you will undertake in the next 30 days.

- Deliverable: Add **Classification** and **Mitigation** columns to your table.

Risk ID	...	RPN	Classification	Mitigation
1				
2				
3				

Step 4: Create a Stub "Living" Risk Register

Turn your FMEA table into a register you can update. In your project tool or a shared document, record for each risk:

- **Owner** (who is responsible for mitigation)

- **Deadline** (when the mitigation will be completed)

- Deliverable: A mini-register with columns **Risk ID, Owner, and Deadline.**

Risk ID	Owner	Deadline
1		
2		
3		

Step 5: Schedule Your First Review

Block 30 minutes on your calendar exactly four weeks from today to revisit:

- Updated PO/Severity/RPN based on new data

- Progress on your mitigation actions

- Any new hazards to add

- Deliverable: A calendar invite titled "Risk Register Review" with an agenda note.

Bonus: Reflect on Your Risk Management Maturity

On a scale from 1 (novice) to 5 (expert), how confident are you that:

- **You've identified the right hazards?**

- **Your PO/Severity scores reflect reality?**

- **Your mitigations will move the needle?**

Write down one area you'll focus on improving before your next review.

Completing this challenge means you've converted abstract fears into ranked, actionable tasks, and you've built the framework to keep reducing risk as you go.

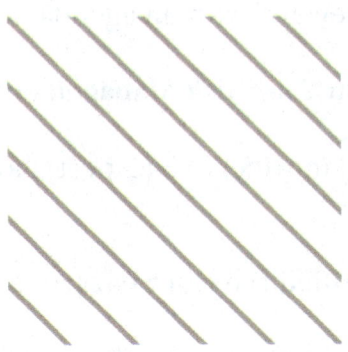

Part IX

Structuring For Success

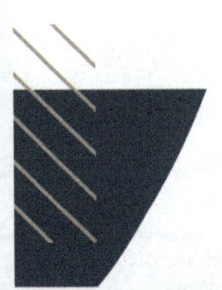

CHAPTER 25: THE LEAN CANVAS AS A LIVING TOOL

The Lean Canvas is not a worksheet. It's a control panel for your venture.

When used right, it helps you clarify strategy, track insights, communicate your model clearly, and keep yourself honest. But most importantly, it's a tool that will be *asked for*.

Why It Matters for Fundraising

Whether you're applying to Y Combinator, joining an accelerator, or pitching a VC, you'll almost certainly be asked: **"Can you share your Lean Canvas?"**

Accelerators use it to quickly understand how well you've mapped the market.
VCs use it to judge clarity of thinking, not just product design.
Internal stakeholders use it to track changes and stay aligned.

Done well, it becomes a **summary of what you know**, what you've tested, and what you still need to learn. Done poorly, it's just another forgotten doc in your drive.

1. THE LEAN CANVAS: BOX-BY-BOX BREAKDOWN

Let's get very clear on what each part of the Lean Canvas means. These are the nine boxes and what you should be trying to accomplish in each.

Box	What It Means
Problem	What key problems are you solving? These should be deep, real, and ideally urgent. Think from the user's perspective, not your product's.
Customer Segments	Who exactly are you solving for? Be specific. "Hospitals" is not a segment; "wound care teams in regional hospitals under 500 beds" is.
Unique Value Prop	What measurable benefit will the customer or end user receive? This should be a concise articulation of the impact you're delivering, not how you deliver it.
Solution	What mechanism, tool, or approach are you using to achieve the Unique Value Prop? Think of this as the how, not the why.
Channels	How will you reach your customers? What are the marketing, referral, or distribution strategies to get traction?
Revenue Streams	How will you make money? This could be per-user licensing, reimbursement, partnerships, direct sales, etc.
Cost Structure	What will it cost to run this venture? Think in terms of key recurring expenses (not every item, but the big ones).
Key Metrics	What numbers will tell you whether you're succeeding? Not vanity metrics… choose ones that drive decision-making.
Unfair Advantage	What do you have that others don't and can't easily replicate? (e.g., data, partnerships, insider access, IP, reputation, distribution)

※ **BMTT Tip:** Don't stress if your "Unfair Advantage" is empty early on. The goal is to build toward one as your startup matures.

2. EACH BOX SHOULD LINK TO REAL WORK

Once you've written strong, honest answers, the next step is making the Lean Canvas an interface to your actual learning and experiments.

Here's how to treat each box as a clickable map to the work behind your assumptions:

Box	Tie-In
Problem	Link to user interviews, observation notes, and quotes
Customer Segments	Link to persona documents and stakeholder maps
Unique Value Prop	Link to language tested in cold outreach, landing pages, or demos
Solution	Link to sketches, prototypes, or solution validation notes
Channels	Link to marketing tests, email campaign data, or community engagement logs
Revenue Streams	Link to pricing experiments, grant wins, LOIs, or reimbursement research
Cost Structure	Link to early financial models or unit economics calculations
Key Metrics	Link to traction dashboards or learning milestones
Unfair Advantage	Link to IP filings, exclusive data access, advisor leverage, or networks

❄ **BMTT Tip:** Structure your Lean Canvas as a Google Doc, Notion page, or other format that supports **internal linking**. Page 1: visual canvas. Each section links to a breakdown page full of evolving work.

3. INTERNAL VS. EXTERNAL VERSIONS

Internal Canvas
Your internal version is messy, honest, and built for you. Use this to track assumptions, mark risks, and log insights over time. Annotate freely: "Still testing this," "Stakeholder response unclear," or "Refining this box next."

External Canvas
Your external version is polished, readable, and sharp. Strip out the commentary. Use this version for advisors, pitch decks, or grant applications. It's your story, clean and digestible.

Mistake to Avoid: Don't use your external canvas as your working document. If your team can't speak openly in the canvas, you're using it wrong.

4. MAKE IT A LIVING DOCUMENT

Set up your canvas so it evolves. It should not live in PowerPoint or as a locked PDF.

Setup Suggestions

- **First Page** = Lean Canvas visual
- **Each Box** = Clickable link to deeper page
- **Version History** = Track monthly changes

- **Quarterly Review** = Scheduled check-ins

- **New Hires** = This is the first doc they should read

If you're feeling ambitious, create mini-canvases for specific stakeholder segments (e.g., "Canvas for Hospital Admins," "Canvas for Home Health Nurses"). This helps you understand how your model resonates differently with each group.

5. COMMON MISTAKES AND HOW TO FIX THEM

Mistake	Fix
Writing generic customer segments	Get specific. You're not targeting "hospitals", you're targeting roles.
Treating assumptions as facts	Link to validation efforts (surveys, interviews, tests)
Using buzzwords like "revolutionary" or "AI-driven"	Write like you're explaining to a teammate, not pitching an investor
Letting the canvas stagnate	Use your calendar. Revisit it monthly
Ignoring risk or weak spots	Mark and prioritize the riskiest box

6. SAMPLE LEAN CANVAS (CLINICAL EXAMPLE)

Box	Example Entry
Problem	Post-surgical infections often go undetected until they become serious
Customer Segments	Outpatient surgical patients and home health nurses
Unique Value Prop	Reduce post-surgical infection prevalence by 34% through early detection
Solution	Smartphone app for guided photo uploads, symptom tracking, and clinician alerts
Channels	Hospital discharge partnerships and nurse training networks
Revenue Streams	Per-patient provider licenses or bundled discharge service contracts
Cost Structure	App maintenance, AI model tuning, compliance and regulatory upkeep
Key Metrics	Infection detection rate, # daily uploads, provider retention
Unfair Advantage	Exclusive pilot program data with 3 hospitals and a closed-loop feedback system from nursing staff

CASE REFLECTION: THE CANVAS THAT ACTUALLY CHANGED THE STARTUP

1. The Team That Treated the Canvas Like a Checklist and Missed Their Blind Spot

A student-led healthtech venture building a vitals-monitoring wearable had an early Lean Canvas that looked polished but never changed. Despite running clinical interviews, usability tests, and early price experiments, they never updated the "Customer Segments" box. They assumed hospital systems were their buyers. Six months in, they realized procurement timelines were a year out, and budget authority was fragmented. It was actually outpatient rehab clinics that moved faster and saw immediate value.

Lesson: A Lean Canvas is only useful if it evolves with what you learn. Static = stale = risky.

2. The Founder Who Used Their Canvas as a Tactical Map

A solo founder targeting eldercare used her Lean Canvas as a personal compass. Every time she talked to a new stakeholder (caregivers, nurses, adult children), she annotated boxes, refined assumptions, and linked in quotes or feedback. When applying to an accelerator, she shared both her "polished" version and a version showing tracked changes over 6 months. The reviewers told her afterward: *"Your clarity of thinking stood out instantly."*

Lesson: The Lean Canvas isn't just for clarity; it's a credibility tool when treated like a living journal.

FINAL THOUGHT

The Lean Canvas is not just for pre-product startups. It's a reflection of how well you understand your market, your user, and your path.

Every time something major changes (your GTM strategy, your customer segment, your solution, etc.), your Lean Canvas should change too. If it doesn't, that's a signal you're building in the wrong direction.

Done well, your Lean Canvas is not just a fundraising document. It's your startup's operating compass.

CHALLENGE: Build Your Live Lean Canvas

This chapter wasn't theoretical; it was a direct invitation to action. Now it's your turn.

Your challenge is to **create your Lean Canvas as a live, evolving document** that you'll return to regularly, not just build once and forget.

Step 1: Choose Your Format

Pick a platform where you can:

- View all 9 boxes at once
- Link each box to deeper documentation
- Revisit and revise over time

Recommended tools:

- **Notion**
- **Google Docs/Sheets with internal links**
- **Miro or FigJam** for a visual front-end, paired with a shared folder
- **Leanstack** if you want a plug-and-play canvas builder

Step 2: Fill In All 9 Boxes

Be honest, not perfect. If something's still unknown, *mark it that way* and add a next step to validate.

1. **Problem**
2. **Customer Segments**
3. **Unique Value Proposition**
4. **Solution**
5. **Channels**
6. **Revenue Streams**
7. **Cost Structure**
8. **Key Metrics**
9. **Unfair Advantage**

Bonus: Under each box, add 1–2 bullet points linking to proof (e.g., user quotes, early tests, pilot conversations, pricing data, sketches).

Step 3: Make It a Living Document

This is not a one-and-done template. Ideally, you would generate at least a page of supporting details for each of the 9 segments.

Set calendar reminders to:

- Review it monthly (or after major pivots)

- Share it with co-founders, advisors, and early hires

- Track version history (via comments or dated snapshots)

Optional Stretch Goal

If your startup serves multiple stakeholders (e.g., patients and clinicians), create a **stakeholder-specific Lean Canvas** for each. This helps clarify which boxes change depending on who you're solving for.

Anchor This

Startups don't fail because of a single bad slide or missed meeting. They fail because they build without clarity, chase without strategy, and grow without alignment.

Your Lean Canvas is not a worksheet. It's your **startup's cockpit**. Build it once, then fly with it, checking your instruments as you go.

CHAPTER 26: PITCHING WITH PURPOSE

You will receive more advice on pitch decks than any other part of your startup. Founders obsess over what to say, what to show, what to wear, when to breathe. And in all the noise, they often miss the only things that truly matter:

- Your slides are clear, sharp, and beautiful.
- Your grammar is perfect.
- Your word count is minimal.
- You present as a genuine expert in your subject.

Everything else is a bonus.

This chapter will help you do exactly that. Whether you're pitching to angel investors, venture firms, competition judges, or accelerator programs, you'll leave with a deck that's more than just "fine"; you'll leave with a deck that gets remembered.

THE BARE MINIMUM: SLIDE HYGIENE

Before worrying about structure, story arcs, or investor psychology, lock in the fundamentals. A weak deck can kill interest before you've even opened your mouth.

Slide Hygiene Checklist:

- Minimal word count (no full paragraphs… ever).
- Fonts: consistent and legible (no comic sans, ever).
- Clean color palette with sufficient contrast.

- No typos or grammatical errors. Zero.

- Graphs and visuals, not text walls.

- Use icons or diagrams for flow, not decoration.

❋ **BMTT Tip:** Pay a designer if you need to. A $500 pitch design is often more valuable than your first patent filing. Investors can't read minds, but they will assume messy slides reflect a messy company.

Slide Structure That Follows Investor Logic

Although every investor is different, nearly all think in a linear, de-risking framework. A basic pitch structure that aligns with how they evaluate you might look like this:

Core Structure:

1. **The Problem**
   What is broken and who suffers because of it? Be specific, be human, be visual.

2. **The Solution**
   What are you building? What does it look like? Show the product; do not just describe it.

3. **Unique Value Proposition**
   What benefit does your solution provide that no one else can match?

4. **Market**
   Who pays for this? How big is that opportunity, and how do you know?

5. **Traction**
   What have you proven so far? Pilots, users, partnerships, LOIs, etc.

6. **Business Model**
   How do you make money or deliver value?

7. **Go-to-Market Strategy**
   How will you find customers and grow?

8. **Team**
   Why you? What unique mix of skills makes your team the one to solve this?

9. **Ask**
   How much are you raising, what is it for, and what will it unlock?

Optional:

- **Why now?** (Timing tailwinds)

- **Competitive landscape** (Use a 2x2 matrix or value map, not a text list)

NARRATIVE ARCS THAT WORK

Investors are humans. And like all humans, they think in stories. If your deck is the *map*, your verbal pitch is the *journey*.

Narrative Flow That Resonates:

- Start with the human pain.

- Show the breakthrough insight.

- Map that insight into a credible, fundable product.

- Show traction, early wins, or proof that you're not guessing.

- Close with ambition and realism: "Here's the vision, here's how we get there."

WHAT COMES AFTER: FOLLOW-UPS AND ETIQUETTE

Live Demo Dos and Don'ts

- **DO** rehearse it like a stage play.

- **DO** test your tech across different devices and Wi-Fi speeds.

- **DON'T** rely on live data. Simulate where needed.

- **DON'T** stall or panic if something fails: pivot smoothly.

Investor Follow-Ups:

- Always send your deck *as a* PDF, not a PowerPoint file.

- Include a 1-pager and Lean Canvas if requested.

- Thank them. Briefly recap next steps.

- Be prompt, be polished, be gracious.

BMTT BREAKDOWN: TAILORING THE PITCH TO THE AUDIENCE

Angels

- They often invest emotionally.
- Highlight human stories and community impact.
- Be warm, confident, and straightforward.

Venture Capitalists

- They think in terms of scale, exits, and upside.
- Emphasize defensibility, market expansion, and potential returns.

Judges (Competitions)

- They reward clarity, confidence, and insight.
- Time yourself perfectly. Never go over.

CASE REFLECTION: THE DECK THAT CLOSED THE ROOM

1. The Overbuilt Deck with Zero Impact

An early-stage medtech startup entered a major competition with a complex slide deck: 18 slides, dense bullet points, multiple fonts, and three full paragraphs explaining their regulatory strategy. The science was solid, but the judges disengaged halfway through. Post-event feedback? "We couldn't tell what they actually did, and it felt like a homework assignment." They didn't make the finals.

Lesson: Subject-matter expertise means nothing if you can't communicate it clearly. The best decks earn attention fast and guide it with precision.

2. The Founder Who Got a Yes from a Single Slide

At a pitch night for healthcare angels, a solo founder used a deck with almost no text. One slide showed a photo of a diabetic wound and read, "This is what we're trying to stop." The room went quiet. She followed with a clear, confident pitch, tight visuals, and a smooth 6-minute delivery. By the end, three investors asked for a follow-up, despite the founder being pre-revenue.

Lesson: Investors respond to clarity, confidence, and visual focus. If your first three slides don't hit, the rest might not even be seen.

FINAL WORD: YOU ARE THE PITCH

This can't be stressed enough: your confidence, clarity, and command of your material are more important than any slide sequence.

If your pitch:

- Is aesthetically tight
- Demonstrates irrefutable subject-matter expertise
- Shows obvious command over the risk landscape

Then you are 80% of the way there.

CHALLENGE: Craft Your Investor-Ready Pitch Deck

You've refined the insight. Now it's time to communicate it.

Your challenge is to create a **10-slide pitch deck** that speaks with clarity, confidence, and visual sharpness. This isn't just about looking good; it's about sounding fundable.

Step 1: Follow the Structure That Investors Expect

Use this tried-and-true 10-slide framework:

1. Problem
2. Solution
3. Unique Value Proposition
4. Market Size & Opportunity
5. Traction
6. Business Model
7. Go-to-Market Strategy
8. Team
9. The Ask
10. (Optional) Competitive Landscape or "Why Now?"

Keep it crisp. Keep it visual. No full paragraphs EVER.

Step 2: Design for Legibility and Impact

Use any modern presentation platform (e.g., PowerPoint, Keynote, Canva, Pitch, etc.) and follow these best practices:

- One message per slide
- Consistent fonts and color palette
- High contrast for readability
- No typos, no clutter
- Visuals and graphs over blocks of text

Tip: If design isn't your strength, get help. A clean, beautiful pitch deck often opens more doors than your first clinical pilot.

Step 3: Critique and Refine

Before sharing your deck, test it with at least one of the following:

- A peer or advisor who understands your domain
- A trusted mentor or investor contact
- An AI-powered critique tool (many now offer pitch deck analysis, tone reviews, and pacing suggestions)

There are dozens of AI tools available that can review decks, suggest improvements, or simulate audience reactions. Use them if they add value, but don't rely on them to write your pitch for you. You are the expert.

Step 4: Rehearse Your Delivery

Practice your verbal pitch until you can:

- Finish in **under 5 minutes**

- Speak naturally and confidently without reading from slides

- Handle questions with grace and clarity

If possible, record yourself and review with a critical eye, or better yet, with someone who'll be honest.

Optional Stretch Goal

Create two versions of your deck:

- A **live pitch version** with minimal text, built for storytelling

- A **standalone version** with enough clarity to be shared by email

Anchor This

Your pitch deck is not a summary of your idea. It's a signal of your clarity, readiness, and respect for your audience's time. Build it like you'd build your product: beautifully, deliberately, and with the user (your investor) in mind.

If you want to see strong examples, look at collections like **Sequoia Capital's "Writing a Business Plan" deck**, the **Y Combinator library of pitch resources**, or slide libraries shared by VCs and founders on platforms like **Pitch.com** and **LinkedIn**. Many founders also publish their successful decks on **Medium** or **SlideShare**, often annotated with what worked and what didn't. These can help you compare your slides to real-world fundraising material.

CHAPTER 27: BUILDING A TEAM AND LEADING IT

Startups are built by people, not pitch decks.

You will pivot. You will run out of money. You will rewrite your roadmap at midnight after a failed pilot. Through all of it, your company will stand (or fall) based on who is next to you when it happens.

This chapter is about the real backbone of startup success: the people you bring in, the culture you create, and the way you lead under uncertainty.

HIRE FOR THE STORM SHELTER

Don't build a startup with someone just because they check a box.

Your co-founder and first few hires will shape everything: your strategy, your stamina, your sanity. So before we talk about resumes and roles, here's a filter that matters more:
Would you want to be in a storm shelter with this person during a hurricane?

Early-stage startup life is chaos. Choose people you can communicate with under pressure, people who don't flinch when things go wrong, people who aren't afraid to challenge you. Skillsets matter, but chemistry and conflict management matter more.

A word of caution: many students default to co-founding with their closest friends. This can work, but only if both sides are ready for tough conversations about equity, roles, and conflict. If you avoid those conversations to "protect the friendship," the company will suffer, and the friendship might too.

Also consider complementarity. Some of the strongest co-founding teams are built on differences: one technical, the other more outward-facing and charismatic; one a builder, the other a storyteller. If you are not naturally a people person, recognize that investors, partners, and early hires will still expect warmth and likability from someone at the table. Sometimes the right co-founder is the person who balances you out in those ways.

BMTT Tip: Avoid co-founding with people who agree with everything you say. If everyone has the same instincts, you're not covering enough ground.

HIRING YOUR FIRST KEY ROLES: WHAT REALLY MATTERS

Your first hires are not just filling roles, they are shaping the company's DNA. Here's what to prioritize:

1. Raw Learning Speed

Early hires must solve problems no one has solved before. The best trait you can screen for is the ability to *learn fast and autonomously*.

2. Bias Toward Action

In an early-stage startup, there's no room for paralysis. Look for people who default to trying something small, even if it's wrong, over endless theorizing.

3. Ownership Mentality

You need teammates who say *"this broke, so I fixed it,"* not *"that's not my job."*

4. Complementary Skills

You don't need five builders or five closers. Think like a systems engineer: what gaps need filling?

Do You Actually Need a Co-Founder?
Many founders assume that they must find a co-founder before they can start. In reality, some thrive as solo founders with a strong technical or operational team around them. The decision depends on:

Skill Coverage: If you are missing a critical function (for example, you are a scientist with no one to handle business development), a co-founder may make sense. If you can hire or contract that expertise, you may not need to give away founder equity.

Temperament: Some people thrive on collaborative decision-making; others prefer to lead decisively with input from a team. Be honest about your style.

Investor Expectations: In some sectors, investors prefer teams over solo founders, but what they truly want is confidence that all core skills are covered. A solo founder with a strong leadership team can be just as fundable as a co-founded venture.

Trust and Commitment: Co-founders are not just early hires; they are equity partners for years to come. If you are not ready to tie your company's future to another person at that level, it is better to remain solo than to partner prematurely.

Funding Realities: Early-stage building often means zero or lean operational capital, which pushes many founders to bring on partners instead of paying hires. While equity-based partnerships can work, do not let lack of cash force you into a permanent co-founder relationship. There are always lean ways to start moving, such as no-code prototypes, student collaborations, part-time contractors, or grant funding. Treat equity as the most expensive currency you have.

⁂ **BMTT Tip:** If you are debating between a co-founder and a first key hire, ask yourself: *"Do I want this person to be a permanent owner of the company, or do I want them to be an early leader who grows with it?"* The answer often clarifies the right path.

STRATEGIC MENTORS AND ADVISORS

You can hire talent and recruit co-founders, but some of the most valuable people on your journey may never take a paycheck or equity check. They are the ones who offer perspective when you cannot see straight, open doors you could not push yourself, and tell you the truths you do not want to hear but need to. These are mentors and advisors.

Mentors vs. Advisors
Mentors are relationships that help you grow as a founder. They invest in your development, not just your company. They will ask about your burnout, your blind spots, your leadership style.

Advisors are relationships that help your startup grow. They bring targeted expertise in regulation, fundraising, clinical trials, or go-to-market. They might join a formal advisory board and hold equity, or they may just be a trusted voice on call when you face a decision in their domain.

You need both. Mentors shape you. Advisors shape the company.

How to Attract Mentors and Advisors
The default assumption is that students cannot "offer value" to high-caliber people. That is not true. Your student status is often your greatest asset. Many accomplished people are eager to

invest in the next generation, but you have to approach them with respect and clarity.

Leverage Student Status
Do not apologize for being a student; use it. A cold email that begins, "I am a student trying to build my first venture and I would love your advice on X..." is disarming. Many leaders will give you 15 minutes just because you are still in school.

Research Before Reaching Out
Never ask questions you could answer with a five-minute Google search. Instead of, "Can you tell me about your career?" try, "I saw your FDA commentary on device regulation and I am struggling to navigate the same issue. Could you share how you approached it?" That specificity earns respect.

Be Clear About Next Steps
Mentors are not mind readers. If you are wrestling with whether to raise capital, or how to validate a clinical need, say so. Define your assumed next step and your current obstacle. Then invite their perspective: "My plan is to run 20 more customer interviews before prototyping, but I am unsure if I should bring a clinician into the team first. What do you think?"

Follow Up Thoughtfully
Nothing kills a budding mentor relationship faster than silence. If someone gives you advice, update them on what you did with it. "You suggested I test assumptions with short surveys before scheduling long interviews. I did, and it helped me refine my target customer profile. Thank you." This builds trust and makes them more likely to keep investing in you.

Offer Value Back
Even as a student, you can give. Share a relevant article, introduce them to a classmate working on something interesting, or offer

your perspective on student markets. Mentorship is not one-way charity; it is a relationship.

Formalizing Advisors
At some point, your company will benefit from formal advisors. This does not have to be complicated:

- Create a small advisory board (2–4 people).

- Offer modest equity (0.25%–1%) for a 1–2 year term, with quarterly calls and defined expectations.

- Keep the bar high. One thoughtful, engaged advisor is better than five names on a slide deck who never answer your email.

BMTT Tip: Do not chase mentors or advisors for their brand name. A "celebrity advisor" who never shows up is worth less than a mid-career professional who responds quickly and challenges your thinking.

Case Reflection: When Mentorship Accelerated Growth
A student-led team in digital health struggled to understand hospital purchasing. They reached out to a retired hospital administrator who had never heard of their university but agreed to meet because they were students. Within weeks, he connected them to three buyers, helped them cut six months of guesswork, and ultimately joined their advisory board. That relationship did not just accelerate the company, it accelerated the students' growth as founders.

LEADING THROUGH AMBIGUITY AND SHAPING EARLY CULTURE

Leadership in the early stage is not about giving answers, it's about creating clarity when none exists.

What founders often get wrong:

- Waiting too long to set expectations ("we're still figuring it out" is not a strategy).

- Confusing transparency with dumping stress on the team.

- Assuming culture will emerge naturally.

BMTT Rule: Culture is the byproduct of your defaults.

- If you always debug problems with the team present, your culture becomes collaborative.

- If you celebrate small wins publicly, your culture rewards progress.

- If you never talk about burnout, your culture silences it.

Shape it early. Live it daily.

FOUNDER PSYCHOLOGY: BURNOUT, CONFLICT, AND DELEGATION

Being a founder means you are always at risk of giving too much. You'll wake up thinking about it. You'll go to bed still thinking about it. And the pressure to "do it all" is real.

Here are some grounding reminders:

1. Burnout Doesn't Look Like Weakness

It often looks like overcommitment, irritability, or hyper-productivity. Take breaks before you need them.

2. Conflict Will Happen

Avoiding conflict is not the same as resolving it. Practice non-defensive communication. Name tensions early. Default to directness.

3. Learn to Delegate (Yes, Even When You're Better at It)

Yes, you might do the task faster. But leadership means enabling others to grow. Your job isn't to be the best at everything. It's to build the best team around you.

Optional Tip: When to Give Titles, and When Not To

Early titles can be helpful or harmful. Here's a guide:

When to delay titles:

- If your team is still exploring what roles they're best suited for.

- If you don't yet have a clear org structure and responsibilities.

When titles help:

- When engaging external stakeholders who need to know who's leading what.

- When it adds legitimacy to investor or partner conversations.

Rule of Thumb: Titles should reflect responsibility and clarity, not ego or assumptions.

CASE REFLECTION: WHEN THE CO-FOUNDER FIT WAS WRONG

Case 1:
A promising diagnostics startup launched from a top-tier university lab. The two co-founders were technically brilliant, had secured non-dilutive funding, and won multiple pitch competitions. But behind the scenes, decision-making stalled constantly. One wanted to move fast and pivot based on feedback; the other clung tightly to academic rigor and perfectionism. They never resolved the conflict: just avoided it.

Six months in, the company missed key deadlines, their lead engineer left, and one co-founder quietly exited. It took nearly a year to rebuild momentum.

Lesson:
Skill isn't enough. If you can't resolve conflict constructively or disagree without derailing progress, even the best idea can stall. Founding team alignment isn't soft; it's structural.

Case 2:
A wearable device startup raised a strong pre-seed round and hired three interns to support R&D. The founding team never articulated expectations for documentation, feedback, or how to handle setbacks. When the first prototype failed bench testing, no one told the founders until three weeks later.

By then, deadlines had slipped, morale was low, and interns were confused about their roles. The founders weren't neglectful; they were just too busy to realize their silence had become the culture.

Lesson:
Culture forms whether you shape it or not. If you don't define how your team works, communicates, and handles failure, the default will fill in the gaps and often not in your favor.

CLOSING THOUGHT

You will make hundreds of decisions in your startup's first year. None matter more than who you bring into the room. Prioritize resilience. Prioritize challenge. Prioritize chemistry.

And when you find the right people, invest in them like your company depends on it, because it does.

HANDS-ON ASSESSMENTS AND TOOLS

To help you apply these lessons, use the following practical frameworks and checklists:

1. Founding Team Compatibility Diagnostic

Rate yourself and your co-founder(s) on a scale from 1 (not at all) to 5 (absolutely true).

Question	You	Co-Founder
We can resolve disagreements without outside help.		
We have complementary skills and split responsibilities clearly.		
We've handled stress together (school, crisis, project).		
We can be radically honest with each other.		
I would still trust this person with money or authority if I left the company.		

※ **BMTT Tip:** A score under 18 suggests you should set up formal conflict resolution agreements and define ownership contingencies before scaling.

2. Equity Split Exercise: Sweat, Role, and Risk

Use this basic rubric to discuss equity fairly.

Category	Weight	You (%)	Co-Founder (%)
Role criticality	30%		
Time committed	25%		
Financial risk	10%		
IP or asset contribution	15%		
Initiative and founding effort	20%		

Use the output as a launchpad for a fair, informed equity discussion.

3. Culture Kickstart Template

- Three things we always do:
- Three things we never do:
- How we give and receive feedback:
- How we celebrate:
- When we prioritize speed over quality (or vice versa):

✕ **Example:** A remote-first medtech team set a "we always over-communicate assumptions" rule to avoid silent misalignment during async work.

4. Delegation Heatmap

Identify what to delegate.

Task	Done By	Skill Level (1–5)	Want to Keep? (Y/N)	Delegate Now?
Social Media	Founder A	2	N	✓
Financial Modeling	Founder B	5	Y	✕
Prototype Assembly	Intern	3	N	⚠ (needs oversight)

5. Founder Wellness Self-Check

Answer weekly. Three or more "Yes" responses? Pause and reevaluate.

- Am I sleeping fewer than 6 hours per night?
- Did I avoid checking in with my co-founder this week?
- Am I taking on tasks I resent out of fear that others will do them worse?
- Do I feel detached from the company's progress?

Treat this chapter as a reflection mirror *and* a toolbox. The best founders don't just build products; they build teams that make building sustainable.

CHALLENGE: Build the Team That Builds the Startup

The success of your venture depends less on the quality of your product and more on the quality of the people building it with you.

This challenge invites you to take an honest look at your team, your leadership habits, and the early culture you're creating before growth locks in patterns that are hard to undo.

Step 1: Run the Founding Team Diagnostic

Download or recreate the **Founding Team Compatibility Diagnostic**. Fill it out yourself, and if possible, have your co-founder(s) do the same. Compare answers.

If your combined score is below 18:

- Schedule a structured alignment conversation.

- Draft a basic co-founder agreement that covers conflict resolution, decision-making rights, and exit contingencies.

Step 2: Equity Alignment Exercise

Use the **Sweat, Role, and Risk rubric** provided in this chapter to estimate a fair equity split. Fill in the table together. Use this as a neutral starting point to guide an open discussion.

Important: This isn't a binding outcome; it's a conversation tool. But it will surface assumptions you need to address now, not later.

Step 3: Draft Your Culture Kickstart Template

With your current team (or even solo if pre-team), fill out these five prompts:

- Three things we always do:
- Three things we never do:
- How we give and receive feedback:
- How we celebrate:
- When we prioritize speed over quality (or vice versa):

Store it in a living document. Revisit and refine quarterly, especially as new teammates join.

Step 4: Create a Delegation Heatmap

List your top 10 recurring tasks. For each, rate:

- Who's doing it now
- Their skill level (1–5)
- Whether they want to keep doing it
- Whether it should be delegated (now or soon)

Use this to guide handoffs and identify early hiring or intern needs. Delegation is not a luxury; it's a necessity.

Step 5: Map Your Mentor and Advisor Needs

List the top four gaps that slow you down today. For each gap, label it as founder growth or company growth.

- Founder growth needs point to **mentors**. Examples: leadership under pressure, time management, hiring, investor communication.

- Company growth needs point to **advisors**. Examples: FDA pathway, clinical study design, hospital procurement, enterprise sales.

Create two short lists:

- Two potential mentors who can help you grow as a leader.

- Two potential advisors who can help the company cross a technical or commercial hurdle.

Mentor Fit Scorecard (0 to 5 each, total out of 25)

- Experience relevant to my growth area

- Communication style fit

- Availability for a monthly check-in

- Track record of developing younger leaders

- No conflicting interests

Score 20 or above: strong fit. Score 15 to 19: good fit with expectations set. Below 15: meet for advice first, do not formalize yet.

Advisor Fit Scorecard (0 to 5 each, total out of 30)

- Domain relevance to our immediate milestones
- Access to buyers, pilots, or data we need
- Willingness to engage on a defined cadence
- Reputation and credibility in our market
- No competitive conflict
- Practicality and bias toward action

Score 24 or above: green light to formalize. Score 18 to 23: start as informal advisor and reassess. Below 18: keep as a friendly expert.

Step 6: Draft Two Outreach Notes and Send One

Write two short emails you can copy and adapt.

Mentor outreach template

Subject: Student founder seeking guidance on [specific growth area]

Hi [Name],

I am a student founder at [school or program] working on [one-line venture]. I have been learning from your work on [specific item you read or watched]. I am working through [one sentence on challenge].

Would you be open to a short call next week to get your perspective on [two precise questions]? I can send a one-pager before the call.

Thank you,

[Name]

[Role]

[Link]

Advisor outreach template

Subject: Quick diligence question on [domain] for [startup name]

Hi [Name],

I am building [one-line venture]. We are preparing for [specific milestone, for example a pre-sub or pilot]. Your experience with [specific company or paper] is directly relevant.

Could I get 20 minutes to validate our approach to [one decision]? I will share a one-page summary with our current plan and the exact question.

Thank you,

[Name]

[Link]

Send one of these today. Put a calendar reminder to follow up in one week if you do not hear back.

Step 7: Prepare a One-Page Advisor Brief

If an advisor expresses interest, share a single page that covers:

- Problem and wedge in two sentences
- Your next two milestones and dates
- The decision you are asking them to help with
- Expected cadence, for example one call per quarter, and one email update per month
- Draft term for formal advisors, for example 0.25 to 1 percent equity, one to two year term, quarterly calls, observer only
- Keep it simple and specific. Invite edits.

Optional Stretch Goal: Founder Wellness Self-Check

Use the 5-question self-check weekly for one month. Don't skip it. If you answer "yes" to three or more, that's your cue to pause, talk to someone you trust, and course-correct.

Anchor This

Your startup won't survive because your pitch deck is flawless. It will survive because your team shows up even when the plan doesn't.

So choose wisely. Lead deliberately. And remember: the company you build is only as strong as the people who believe in it enough to build it with you.

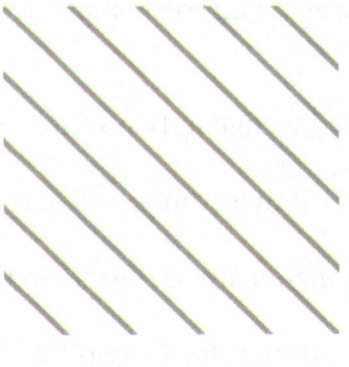

Part X

Financial Fluency For Founders

CHAPTER 28: DEMYSTIFYING FUNDING PATHWAYS

Money is Not Just Money

You need funding. But not all dollars come with the same expectations or consequences.

In this chapter, we unpack the six most common funding pathways for early-stage founders: **bootstrapping, grants, fellowships, angel investment, venture capital, and other non-dilutive options**. Each of these sources gives you money, yes, but each also "wants" something in return.

What they want could be equity, progress, data, goodwill, PR, or nothing at all. But if you don't understand these expectations, you'll misalign your company's structure, timeline, and strategy from Day 1.

1. BOOTSTRAPPING: BUILDING WITH YOUR OWN MONEY (OR TIME)

What it is:
Funding your company with your own capital, labor, or personal network (no outside money).

What it wants in return:
Survival. Scrappiness. Progress with zero strings.

Upsides:

- You retain 100% control and ownership
- Forces creative, lean execution
- Keeps expectations realistic

Watch-outs:

- Limited runway and scalability
- Can lead to burnout if sustained too long
- Can delay key experiments that need upfront capital

Use when:
You're still validating core assumptions and don't want to give away equity or aren't ready to ask for it.

2. GRANTS: NON-DILUTIVE CAPITAL THAT BUYS TIME AND CREDIBILITY

What it is:
Government or foundation-funded support for specific R&D, education, public health, or innovation outcomes (e.g., NIH, NSF, SBIR/STTR).

What it wants in return:
Mission-aligned progress. Publications. Deliverables. In some cases, future commercialization.

Upsides:

- Non-dilutive: you don't give up equity
- Adds credibility, especially in technical fields
- Can fund long-term, risky ideas

Watch-outs:

- Highly competitive and slow (6–12+ months)
- Often requires strict deliverables and admin work

- Can pull you into academic timelines and language

Use when:
You're building tech with measurable impact (health, science, energy, education) and want to delay equity fundraising.

3. FELLOWSHIPS: PERSONAL FUNDING WITH STRATEGIC PERKS

What it is:
Individual support for you as a founder, researcher, or innovator - often includes funding, mentorship, and a platform (e.g., Activate, E14, Thiel, NIH fellowships).

What it wants in return:
Your growth. Your success. Your presence as part of their brand or network.

Upsides:

- Often highly founder-centric
- Can cover cost of living and project expenses
- Opens up valuable networks

Watch-outs:

- Some restrict full-time company work
- Can unintentionally delay company formation
- Limited to 1–2 years max

Use when:
You're early and need breathing room to build, validate, and decide your path forward.

4. ANGEL INVESTMENT: HIGH-TRUST, HIGH-RISK EARLY CAPITAL

What it is:
Equity-based investment from high-net-worth individuals, often before product or revenue.

What it wants in return:
A piece of the upside. Influence. Belief that you'll make something big happen.

Upsides:

- Can move quickly with fewer formalities
- Angels often become advisors or doors to VCs
- Good fit for unconventional timelines

Watch-outs:

- Risk of misaligned expectations
- Difficult to price equity accurately at early stage
- Can complicate cap table if not structured well

Use when:
You've validated your problem and are building out early solutions, especially if you already have a warm network.

5. VENTURE CAPITAL: SCALED GROWTH, AT A COST

What it is:
Institutional capital in exchange for equity that is designed to fuel fast growth and large returns (typically 10x+).

What it wants in return:
A category-defining business. Speed. Exit potential. Clear market dominance.

Upsides:

- Large check sizes

- Access to networks, talent, and follow-on capital

- Can de-risk bold plays (hardware, regulatory)

Watch-outs:

- You're giving up control: eventually, lots of it

- Pressure to grow, not just build

- Not designed for "steady state" or slow-burn models

Use when:
You're ready to scale, have traction, and want to build something venture-backable, not just viable.

6. OTHER NON-DILUTIVE SOURCES: PITCH COMPETITIONS, PRIZE FUNDS, INNOVATION LABS

What it is:
Milestone- or competition-based grants given for innovation, often tied to a pitch, a milestone, or a specific theme.

What it wants in return:
Proof that their program worked. Publicity. Impact aligned with their goals.

Upsides:

- Great starter capital ($5k–$100k prizes)
- Useful for refining pitches and messaging
- Often comes with mentorship and exposure

Watch-outs:

- Can become a distraction from actual building
- Might push you into too-polished, not-validated ideas
- Many are one-time awards, not recurring

Use when:
You're pre-product or early prototype and want visibility, feedback, and a small funding bump.

7. COMPARING WHAT THEY "WANT" FROM YOU

Funding Type	What They Want	Equity?	Speed	Strategic Fit
Bootstrapping	Survival & progress	No	Fast	Great for validation, lean teams
Grants	Progress + deliverables	No	Slow	Great for technical/R&D startups
Fellowships	Founder growth	No	Medium	Ideal for early-career builders
Angel Investment	Long-term upside	Yes	Fast	Best for early-but-promising
Venture Capital	Growth + exit	Yes	Medium	Best for scale-first companies
Non-dilutive (misc.)	Visibility + impact	No	Fast	Perfect for momentum-building

8. MIXING SOURCES (WHEN IT WORKS, WHEN IT DOESN'T)

It works when:

- You sequence well: grants first, then angels or VC

- Each source funds a milestone, not just a runway

- Your team can manage the parallel expectations (e.g., grants + equity + competitions)

It doesn't when:

- You're overextended by compliance or reporting
- You signal conflicting goals (slow science + fast scale)
- Your cap table becomes messy before a lead investor joins

CASE REFLECTION: THE WRONG MONEY, RIGHT IDEA

Talia's AI-powered diagnostic tool had everything going for it: a promising prototype, a clinical advisor, and early traction with researchers eager to pilot it. She'd even landed a $10K pitch competition win that helped fund a few key hires. But when a well-known VC fund offered her $1M for 20%, she didn't hesitate.

Three months later, she was burning through cash: **not because she was failing, but because she was scaling the wrong way**. The investors wanted faster patient acquisition, a sales lead, and a roadmap to a Series A. But Talia was still knee-deep in pilot testing and regulatory feedback loops. She needed time, not speed. Proof, not polish.

Her fellowship mentor pulled her aside:
"You didn't raise the wrong amount. You raised from the wrong source for your stage."

The team had skipped over NIH grants because "they take too long," and passed on a founder fellowship because it "felt academic." But now, the very science that made them special was being deprioritized in favor of aggressive metrics that didn't reflect true traction.

The Turning Point
Instead of chasing their next round, they paused. Talia negotiated a slower burn rate, applied for a Phase I SBIR with university

support, and secured a founder-focused fellowship to cover her living costs. It was messy, but she realigned the funding strategy with the company's actual arc.

In hindsight, her mistake wasn't raising money; it was misunderstanding what that money wanted back.

Every check comes with a contract, even when there's no paper.

FINAL WORD

Your funding strategy is more than a timeline; it's a positioning tool. Every dollar you raise is a signal. It tells the world what kind of founder you are, what kind of company you're building, and what your future is expected to look like.

So don't just raise money. Raise the **right** money. At the right time. From the right people.

CHALLENGE: Draft Your Funding Fit Snapshot (Without Making a Decision Yet)

You don't need to know *exactly* how you'll fund your startup. But you *do* need to know what each funding type expects and how that lines up with your current goals, stage, and strengths.

This challenge is about **clarity**, not commitment.

Step 1: List Your Startup's Next 2-3 Milestones

Think in terms of proof, not polish. What are the next meaningful milestones that would increase your credibility, traction, or ability to raise capital?

Examples:

- Complete IRB-approved clinical pilot
- Build functional MVP and get 5 letters of intent
- Submit provisional patent and secure SBIR Phase I

Deliverable: Write down your 2-3 milestones and what each one proves.

Step 2: Evaluate Fit Across Funding Types

Use the grid below to rate each funding source on its **alignment with your current stage**. Don't worry about the future yet; focus only on where you are now.

Funding Type	Could It Fund My Next Milestone? (Y/N)	What Would It Want in Return?	My Comfort Level (1–5)
Bootstrapping			
Grants			
Fellowships			
Angel Investment			
Venture Capital			
Non-Dilutive (Other)			

Deliverable: Fill out the table to clarify what might work and where you still need to learn more.

Step 3: Flag What You Don't Know

Identify one funding source from the list above that:

- Seems promising
- But you don't fully understand

Then write **one specific question** you need to answer before seriously considering it.

Examples:

- "What do I actually need to apply for an NSF SBIR?"
- "How much equity does a typical angel take at this stage?"

- "Can fellowships and university IP coexist?"

Deliverable: Write your funding knowledge gap as a question and commit to finding an answer in the next 7 days.

Step 4 (Optional): Funding Strategy Score

Give yourself 1 point for each item completed above:

Action	Completed? (✓)
Milestone List	
Fit Table	
Open Question Identified	

Score: /3

0–1: Time to dig deeper: funding strategy will shape your whole startup arc.
2–3: You're building the right muscle: keep asking smart questions.

Anchor This

You're not choosing a funding source yet. You're **building fluency** in what each one means. The goal isn't to find the fastest money. It's to find the money that helps you build *right*.

The next chapter will give you the tools to protect your company once the offers start rolling in.

For now, start matching your milestones to your funding map, and leave every option on the table.

CHAPTER 29: HOW TO READ A TERM SHEET

What You Sign Is What You Live

You don't learn nuances of a term sheet after you pitch; you pitch because you're ready for one.

If you walk into a VC conversation without understanding the mechanics of startup funding, you're setting yourself up to give away control, misread intentions, or build your company on terms that quietly crush your long-term outcomes.

This chapter is about arming you with the financial and legal fluency you'll need to **protect your company**, **align incentives**, and **walk into investor meetings with confidence**. Whether you're raising $25K or $2.5M, the terms are the architecture of your startup's future.

1. WHY YOU NEED TO UNDERSTAND TERM SHEETS BEFORE PITCHING

When you pitch, you're not just selling your story; you're signaling your **savviness**.

Investors don't expect you to be a lawyer, but they do expect you to understand:

- What kind of capital you're raising (SAFE, convertible note, priced equity)

- What valuation means and how dilution works

- What rights they're asking for and which ones they don't need

Knowing this doesn't just help you negotiate. It helps you **avoid traps**, **build a clean cap table**, and **protect your role as a founder** long-term.

2. SAFE VS. CONVERTIBLE NOTE VS. PRICED ROUND

Fundraising structures can look confusing at first, but most early-stage deals boil down to three main categories: SAFEs, convertible notes, and priced rounds. Each has its own tradeoffs in speed, complexity, and long-term impact. This section is meant to give you a working vocabulary, not a legal education. By the end, you should be able to recognize which option fits your stage, your investor type, and your appetite for structure. Use these definitions as tools in your kit, so you can focus on building rather than second-guessing legal jargon.

SAFE (Simple Agreement for Future Equity)

- **No valuation now**, converts later based on a future round

- Includes either a **valuation cap** or **discount** (sometimes both)

- Simple, fast, and founder-friendly

Use when: You're raising early and don't want to set a hard valuation yet.

Convertible Note

- Like a SAFE, but technically a **loan** that converts to equity later

- Includes interest and maturity date

- Can be risky if the note isn't converted and becomes debt

Use when: You want more structure than a SAFE, or your investors prefer traditional lending logic.

Priced Round

- Investors **buy shares now** at a defined valuation

- Requires setting up a full **Preferred Stock structure**

- Brings full set of rights: board seats, preferences, pro rata, etc.

Use when: You're raising a large round ($1M+), have traction, and need institutional capital.

Check for emerging standardized docs (KISS, CAM) that blend features of SAFEs and convertible notes.

3. THE BIG FIVE TERMS (AND WHAT THEY ACTUALLY MEAN)

If this feels overwhelming, take a breath. You don't need to master every detail right now. Think of this section as a map: you don't need to know every road by heart, you just need to know the major landmarks so you don't get lost when you're in a negotiation later. The goal is awareness, not perfection. Each term below has tripped up founders before, and by knowing the basics, you'll already be ahead of where most first-time entrepreneurs start.

Valuation

- **What it is:** How much your company is worth before and after investment.

- **Why it matters:** Valuation sets the size of the slice you're giving away.

- **Example**: Pre-money valuation = $4M. New investment = $1M. Post-money valuation = $5M ⇨ investor owns 20%.

- **Tip**: A higher valuation is not always better. If it sets unrealistic expectations, it can make the next raise harder.

Dilution

- **What it is**: When new investors buy in, everyone else's percentage ownership decreases.

- **Why it matters**: Dilution is not automatically bad. If your piece gets smaller but the pie gets bigger, you can still come out ahead.

- **Example**: You own 50% of a $1M company = $500K value. After a round, you own 25% of a $10M company = $2.5M value. That's "good dilution."

- **Tip**: Dilution without company growth is a warning sign. Dilution with growth is the fuel that lets you scale.

Liquidation Preferences

- **What it is**: Terms that decide who gets paid first, and how much, when your company is sold or shuts down.

- **Why it matters**: Preferences can shift millions in outcomes between investors and founders. They can matter as much as valuation.

- **Example**:
  • *Non-participating preference*: Investor chooses between taking their $5M back or converting to common and sharing in the upside, whichever is greater.
  • *Participating preference*: Investor takes their

$5M back first, then also joins the common pool, reducing the founders' share.

- **Tip:** Always model scenarios. A "2x participating" preference means an investor gets twice their money back first, then still participates in the upside.

Pro Rata Rights

- **What it is**: The right of an investor to invest in future rounds to maintain their ownership percentage.

- **Why it matters**: It protects early investors from being diluted too much, but if overextended, it can crowd out new capital.

- **Example**: An angel owns 5% after your seed round. In Series A, they can buy enough shares to stay at 5% rather than dropping to 3%.

- **Tip:** These are usually harmless early on, but don't overpromise them to everyone.

Control Terms

- **What it is**: Clauses that define who makes decisions.

- **Why it matters**: Even if you keep most of the equity, you can lose control if terms aren't balanced.

- **Examples:**
    - Board seats: Giving investors too many seats can limit your say.

- o Protective provisions: Requiring investor approval for key actions like selling the company or issuing new stock.

- o Founder vesting: Investors may ask founders to vest their own equity over time.

- **Tip:** Protect control early. A "bad board" can kill a startup faster than lack of funding.

4. CAP TABLES AND WHY THEY MATTER

If the term sheet is the rulebook, the cap table is the scoreboard. It shows exactly who owns what and how that changes with each raise. Many first-time founders ignore their cap table until investors point out problems, but by then it is often too late. This section will help you see the cap table as a living document that reflects not only ownership, but also trust, fairness, and long-term credibility. A clean cap table makes your company easier to fund, easier to grow, and easier to exit.

Your **cap table** is the living ledger of who owns what.

Early-stage cap table priorities:

- Keep it **clean** (avoid too many small checks with rights)
- Keep it **fair** (allocate equity based on risk and contribution)
- Keep it **defensible** (investors will study it closely)

A good cap table lets you:

- Raise money without giving up control
- Reward key team members appropriately
- Exit cleanly, with minimal surprises

5. CONTROL VS. ECONOMICS

In every deal, there are two buckets of negotiation: economics and control. Economics determines who gets paid how much when things go well. Control determines who makes decisions when things are uncertain or tough. Both matter. Investors will often emphasize one while quietly negotiating for the other. This section shows you how to recognize the trade-offs and decide where you are willing to bend. Understanding this balance early gives you confidence and prevents regret later.

To reiterate, every term sheet is a tug-of-war between:

- **Economics**: Who gets what in success
- **Control**: Who decides what happens along the way

VCs will typically push for:

- Protective provisions
- Board seats
- Liquidation preferences
- Founder vesting
- Anti-dilution clauses

Your job is to:

- Understand what each clause means
- Know which ones are standard vs. aggressive
- Decide what you're willing to concede

✳ **BMTT Tip:** You don't need to fight over every term. But don't blindly accept "standard" either. Have a trusted advisor or attorney review the deal: ideally one who **works with startups**, not just big firms.

6. WHAT YOU CAN PUSH BACK ON

Not every term in a term sheet is set in stone. Some are standard, some are negotiable, and a few are red flags. New founders often assume they must accept everything, but that is not true. Investors expect you to push back on the most aggressive asks, and doing so signals maturity rather than resistance. This section highlights the terms where you should draw a line, and offers practical guidance on when and how to negotiate without derailing the deal.

Term	When to Push Back
2x or Participating Liquidation Pref	Almost always a red flag: ask for 1x non-participating
Multiple Board Seats	You don't need to give up more than 1 early on
Founder Vesting	Reasonable if you're getting big VC money but negotiate cliffs and length
Full Ratchet Anti-Dilution	Very aggressive: try for weighted average instead

7. THE "NO REGRETS" FOUNDER CHECKLIST

By now, the terminology may feel dense, but remember: you do not need to memorize every clause. You only need a framework to check your understanding and flag when you need advice. That is the role of this checklist. Think of it as your compass before signing anything. If you can answer each of these questions honestly and clearly, you are already doing better than most first-time founders. Use this as your "pause and reflect" tool before committing to terms that will shape your company for years.

Do I understand the structure of this deal (SAFE vs. priced)?
Do I know how much of my company I'm giving up, both now and in the future?
Do I know who controls key decisions and why?
Do I know what happens if things go wrong (or right)?
Do I have someone experienced reviewing this with me?

CASE REFLECTION: WHY TERM SHEETS DESERVE MORE THAN A SIGNATURE

WIN: A healthtech founder raised $500K on a SAFE with a fair valuation cap and no hidden control clauses. She had an advisor walk her through the terms, built a clean cap table, and secured future pro rata rights for aligned investors. When a VC came in later, there were no surprises, and she retained majority ownership.

LOSE: A robotics startup raised a $1M priced round without understanding liquidation preferences. The 2x participating clause meant that in a modest $5M exit, their lead investor walked away with over half the proceeds. The founders, despite five years of full-time work, barely made back their salaries.

WIN: A student-led biotech team used a convertible note with a clear maturity date and interest terms, raising early capital while preserving flexibility. They converted during a larger priced round a year later, and early investors stayed involved without drama or restructuring.

LOSE: An excited hardware founder gave two angels board seats and protective provisions in their first $100K raise. When real traction hit, no VC wanted to deal with the crowded governance. The deal died on diligence, and the founder had to spend a year unwinding the early agreement just to try again.

FINAL WORD

A term sheet is not just a contract; it's a **vision of how your startup will grow**.

Read it carefully. Negotiate it respectfully. But most importantly, **understand what you're signing up for** before you even take that pitch meeting.

In the next chapter, we'll break down **how to build your fundraising strategy** not just term-by-term, but milestone-by-milestone. Because a great startup doesn't raise rounds. It raises on purpose.

CHALLENGE: Know What You're Signing Before You Even See It

This challenge is not about negotiating a term sheet yet. It's about **recognizing what one actually means** so when the time comes, you're not learning under pressure.

Step 1: Decode the Key Terms in Plain English

Take the five big terms below and re-write each in your own words, without legal or startup jargon. Treat it like you're explaining it to a friend who's smart but unfamiliar with venture funding.

Term	Your Plain-English Definition
Valuation	
Dilution	
Liquidation Preference	
Pro Rata Rights	
Control Terms	

Deliverable: Write each definition in 1–2 sentences, no buzzwords allowed. If you can't explain it simply, you don't understand it yet.

Step 2: Identify What You'd Want to Keep Control Over

Imagine you're offered $500K tomorrow from a VC. What *one or two things* would you want to keep control over, no matter what?

Examples:

- The ability to hire your early team without board approval

- How and when to pivot the product

- What pace the company scales

Deliverable: List 1–2 things you personally would not want to give up control of, and why they matter to you.

Step 3: Look Up a Real Term Sheet (or Use the BMTT Summary)

Read one sample term sheet online (you can search: "Y Combinator SAFE template" or "Series Seed term sheet example"), or re-read the breakdown in this chapter.

Then answer:

- What confused you most?

- What seemed surprisingly fair?

- What felt like a potential red flag?

Deliverable: Write down one insight, one concern, and one question you still have after reading it.

Step 4 (Optional): Quiz Yourself in 60 Seconds

Set a timer for 60 seconds. Try to answer the following from memory:

1. What's the difference between pre-money and post-money valuation?

2. Why might a 2x participating liquidation preference be dangerous?

3. What does pro rata let investors do?

4. What kind of funding (SAFE, note, priced round) gives investors the most control up front?

Scoring:

- 4/4: Good work!

- 2–3: On the right track

- 0–1: Time to revisit the chapter (before a real investor does)

Anchor This

Reading a term sheet *after* it's handed to you is like reading a playbook *after* kickoff. Smart founders understand the game before they're on the field.

You don't need to memorize legalese. But you *do* need to know what you're trading, what you're protecting, and why it matters.

The next chapter will help you map when and how to raise (term sheet in hand or not). For now, train yourself to see the story *behind* the terms.

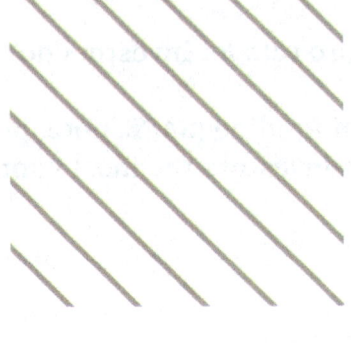

Part XI

Fundraising Strategy and Execution

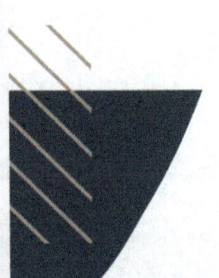

CHAPTER 30: BUILDING A FUNDRAISING STRATEGY

Plan Your Path (Or Follow Someone Else's)

Fundraising is not just about getting money. It's about **getting the right money, at the right time, from the right people**, on terms that support your goals, not derail them.

It's easy to get caught up in the performative grind of pitch competitions, demo days, or Twitter threads full of VC jargon. But capital is only useful when it advances your mission. This chapter is your toolkit for building a **clear, staged fundraising strategy** that supports your actual work, not someone else's expectations.

BMTT Sidebar: Don't Reinvent the Fundraising Wheel

Founders are builders. We're hard-wired to create from scratch, to innovate at every step. But **fundraising is not where you want to blaze a brand-new trail**.

There are thousands of companies, especially in healthcare, deep tech, and mission-driven sectors, that have walked a similar path. Many of their fundraising strategies, timelines, investor types, and round sizes are publicly accessible through:

- Crunchbase
- PitchBook
- SEC Form D filings
- Accelerators' demo day pitch decks
- Your university tech transfer office or incubator network

Use this to your advantage. **Pick 3–5 comparable startups** (by stage, model, or market) and reverse engineer:

- How much they raised
- When they raised it
- From whom
- What they had achieved at each stage

It's not just about copying. It's about grounding your strategy in **investor precedent**, making your pitch more credible, and avoiding painful missteps.

❋ **BMTT Tip:** When speaking to investors, say "We've modeled our plan on similar companies like X and Y, who raised around this stage after reaching similar milestones." That's a story of pattern recognition, not guesswork.

---

1. WHEN TO RAISE: TIMING, TRACTION, AND TRADEOFFS

The right time to raise isn't just when you *can*. It's when:

- You've reached a **clear milestone** that derisks your next phase
- You need capital to accelerate, not to survive
- You've identified the **next 12–18 months** of use for every dollar

Common Early Mistakes:

- Raising money before understanding the market
- Raising too little to reach the next milestone

- Raising too much and losing flexibility or control

- Raising with no plan (or an overly ambitious one)

✻ **BMTT Tip:** If you're asking "How much can I raise?", reframe it as: "How much do I need to unlock the next round of proof?"

2. HOW MUCH TO RAISE: MILESTONE-BASED VS. VANITY-BASED

There's no universal "right amount." But there is a right logic:

Milestone-Based Raise:

- Define **what you need to prove** to get to the next inflection point

- Add buffer for 20–30% runway extension (delays are normal)

- Match capital asks to **hiring, product, clinical, and market needs**

⊗ Example:

- Build functional MVP: $50K

- Run usability pilot + FDA consult: $150K

- Complete preclinical study: $350K

- Total round size: ~$600K with buffer

Vanity-Based Raise:

- "We're raising $1.5M because that's what everyone else raises at this stage"

- No tie to real-world use, burn rate, or validation needs

- Often leads to dilution, bloat, or future failure to hit expectations

3. FROM PRE-SEED TO SERIES A (OR NOT)

Not every path leads to Series A, and that's okay. What matters is building a **coherent progression of validation**:

Stage	Milestones You're Proving	Typical Sources
Pre-Seed	Problem clarity, early users, prototype function	Grants, angel investors, pitch competitions, founders' capital
Seed	MVP traction, early revenue or pilot data, core team	Pre-seed VCs, accelerator follow-ons, super-angels
Series A	Strong traction, revenue or clinical outcomes, repeatable value delivery	Institutional VCs, strategic partners

Some ventures:

- Skip seed and go straight to licensing

- Raise one large seed round and scale

- Bootstrap until revenue sustains growth

There's no shame in taking the **alternative path** if it aligns with your strengths.

※ **BMTT Tip:** A strong fundraising path is one where **each round has a specific job**, and you only raise when that job needs to be done.

4. WHO TO RAISE FROM: ALIGNING CAPITAL WITH MODEL

Different types of capital support different types of business models. Choose funders who understand your space, timeline, and mission.

Fund Source	Best For...	Watch Out For...
NIH/NSF Grants	Science-forward or clinical tech	Long timelines, competitive applications
Angel Investors	Vision-stage validation	Misalignment if expectations aren't clear
Venture Studios	Hands-on co-building	Control and equity trade-offs
Accelerators	Network, feedback, small capital	Distraction if poorly matched
Revenue-Based Investors	Steady income models	Less suitable for deep tech
Strategic Partners	Hardware, clinical, or regulated industries	Longer closing timelines, IP concerns

5. STAGING YOUR FUNDRAISING PLAN

Here's a smart way to **map your path**:

- Step 1: Define Your 3 Key Milestones

These could be:

- Regulatory (e.g., 510(k) submission, IND clearance)
- Technical (e.g., successful prototype, pilot complete)
- Commercial (e.g., signed customer LOIs, first revenue)

- Step 2: Estimate What It Takes to Reach Each

Break down:

- Burn rate (personnel, equipment, outsourcing)
- Timeline (realistic, not best case)
- Buffer (never skip this)

- Step 3: Assign Funding Sources to Each Phase

Align your goals to the **funding stage and source**. Don't try to get venture capital for proof-of-concept. Don't rely on student pitch winnings to fund preclinical trials.

- Step 4: Build a Visual Fundraising Map

Create a slide that shows:

- Phase name (e.g., "Feasibility Phase")

- Milestone(s)
- Timeline (in quarters or months)
- Capital needed and source

Investors love founders with **a plan, not just a pitch.**

6. GOAL-SETTING BASED ON YOUR STARTUP TYPE

Not all traction is revenue.

Model Type	Traction Milestone
SaaS or B2B Tools	MRR growth, customer onboarding time
Clinical Hardware	Preclinical results, regulatory milestones
DTC Consumer Health	User growth, retention, paid conversions
Deep Tech or Drug Dev	Grant wins, research milestones, pipeline IP

Set goals that **investors already associate with progress** in your model. Don't talk revenue if what really matters is a pre-IND meeting or IRB approval.

CASE REFLECTION: THE STARTUP THAT RAISED AT THE WRONG TIME (AND PAID FOR IT TWICE)

A student-founded digital health startup had early traction: a clean prototype, two letters of intent from clinics, and strong university press. Energized by the momentum, the team jumped into a $1M seed raise. They pitched everywhere… accelerators, angels, even some smaller VCs.

Surprisingly, they closed the round in just four months. But they hadn't truly mapped what that $1M needed to prove. There was no regulatory plan, no clinical milestone, and no technical roadmap tied to budget. Within a year, they had burned through $600K trying to "figure it out" with three developers, one BD hire, and a branding agency.

When they tried to raise again, investors asked: "What did you de-risk?" The answer was unclear.

The startup eventually rebooted: this time mapping their milestones: 1) full integration with one clinical partner, 2) a small IRB-backed outcomes pilot, and 3) completion of a HIPAA-compliant backend. They targeted a much smaller raise, just $300K, tied to those three clear outcomes.

This second round didn't just buy them more runway. It bought back investor trust.

Takeaway: It's not how much you raise, it's how clearly that capital connects to a credible next step. Founders don't get extra credit for raising big if they miss the milestones that matter.

FINAL WORD: RAISE WITH PURPOSE, NOT PANIC

Raising money is not a trophy. It's a tool.

And like any tool, it should be used only when it helps you build what matters faster, stronger, or with more reach than you could alone.

Design a fundraising strategy that reflects your timeline, model, and goals. Not what Twitter tells you, not what investors want to hear, **what your company needs to prove next**.

In the next part, we'll cover how to **position your startup**, refine your pitch, and lead your team forward. Because money is fuel. But you're still driving.

CHALLENGE: Map Your Milestones, Not Just Your Money

Too many founders start fundraising without knowing what they're really raising for. This challenge helps you build a fundraising strategy *backwards* by starting with what you need to prove, not what you think you can ask for.

Step 1: Identify Your Next Three Fundraising-Justifying Milestones

List **three specific milestones** that would make your startup more fundable, buildable, or credible. These should be:

- Clear (you'll know when you've hit them)

- Stage-appropriate (don't list FDA approval if you're still prototyping)

- Credible signals to partners, users, or funders

Examples:

- Complete functional MVP with 2 stakeholder test users

- Secure a pre-IND meeting with the FDA

- Obtain 5 Letters of Intent from pilot partners

- Complete IRB-approved pilot study with 10 patients

Milestone	Why It Matters

Step 2: Estimate the Capital Needed to Reach Each Milestone

Don't worry about precision yet: just rough estimates.

Milestone	Capital Needed	Timeline (Months)

Then, add 20–30% as buffer.

Step 3: Match Potential Funding Sources to Each Phase

Using what you learned in previous chapters, match at least one likely funding source to each phase.

Tip: You can mix and match(e.g., grant + competition, or bootstrapping + angel check).

Milestone	Likely Funding Source(s)

Step 4: Create a One-Slide Fundraising Map

(Optional but powerful): Sketch or describe a single-slide fundraising roadmap that shows:

- Each milestone phase (name it!)
- Timeline (Q1, Q2, etc.)
- Capital needs
- Funding source(s)

Investors love seeing clarity. You're not guessing; you're mapping.

Step 5 (Reflect): Choose One Comparable Startup to Reverse Engineer

Pick a startup similar to yours (domain, model, or stage) and use Crunchbase, PitchBook, or online research to answer:

- When did they raise their first round?
- How much did they raise at each stage?
- What milestone did they likely hit to raise it?

Startup Name:

Key Insight You Can Borrow:

Anchor This

A great startup doesn't raise on vibes. It raises on proof.

This challenge isn't about building a perfect budget; it's about connecting your milestones to capital with purpose. Know what you're trying to prove, know what it will cost, and know who funds that kind of progress.

In the next chapter, we'll show you how to unlock non-dilutive money first because smart capital is sometimes free capital.

CHAPTER 31: MONEY WITHOUT DILUTION

Money Without Dilution: Grants, Fellowships, and Bootstrapping

Venture capital is loud. But it's not always the smartest first money in the door. If you're early-stage, unproven, or simply mission-aligned in a way that isn't easily captured by VC metrics, **non-dilutive capital** may be your best (and safest) move.

This chapter walks you through how to access grants, fellowships, competitions, and early scrappy strategies that don't require giving up equity or control: ideal for founders who want to build leverage before raising, or who may never want to raise at all.

1. WHAT NON-DILUTIVE CAPITAL ACTUALLY IS

Non-dilutive funding is any capital that you receive that doesn't require giving up ownership of your company. It includes:

- Government grants (e.g., NIH, NSF, DOE)

- University or institutional fellowships

- Prize money from competitions

- Accelerator or incubator stipends

- Contract revenue from paid pilots or early customers

- Donations or philanthropic funding
  (rare, but sometimes applicable)

It's not "free money"; there are timelines, reporting requirements, and eligibility criteria. But it's a **powerful tool** for building traction, generating IP, and surviving pre-revenue without surrendering your cap table.

2. THE BIG DOGS: NIH, NSF, AND FEDERAL GRANT PATHWAYS

NIH & NSF SBIR/STTR

The **Small Business Innovation Research (SBIR)** and **Small Business Technology Transfer (STTR)** programs are two of the best sources of early-stage capital for scientific and tech-heavy startups.

- **SBIR Phase I**: ~$275K–$300K for feasibility and early proof of concept

- **SBIR Phase II**: ~$1.5M+ for development and early validation

- **STTR**: Similar in scope, but requires a formal partnership with a research institution

Key Agencies to Know:

- NIH (biomedical/health)

- NSF (general science/engineering)

- DOE (energy/climate)

- USDA (agriculture/food)

- DOD (defense)

✳ **BMTT Tip:** Apply as soon as you have a technical objective and a unique angle. Don't wait to be "ready". Phase I is built for risk. There are multiple applications throughout any given year; do not be afraid to fail and get advice or guidance!

What They're Looking For:

- Technical innovation and feasibility
- Societal or commercial impact
- A clear research plan and milestones
- A credible team (doesn't need to be huge!)

3. FELLOWSHIPS THAT FUND FOUNDERS

These aren't just resume lines. Some fellowships come with **serious capital**, lab space, and expert support. Many are surprisingly under-applied to.

Fellowships to Explore:

- **Activate** (formerly Cyclotron Road): ~$400K+ for 2 years, deep tech focus
- **NIH/NIDA Translational Science Fellowships**: Salary, benefits, and NIH mentorship
- **Eisenhower, Thiel, Echoing Green**: For social impact, policy, or alternative education pathways
- **University Innovation Fellowships**: Some include stipends, prototyping funds, or travel grants

※ **BMTT Tip:** Many founder fellowships allow you to keep your startup equity, IP, and commercial rights. Read the fine print, but don't assume it's either/or.

4. STUDENT COMPETITIONS AND LOW-HANGING FRUIT

If you're still a student (or can collaborate with one) **milk every resource your university has.**

Capital Sources to Leverage:

- Business plan competitions (some with $5K–$100K+ prizes)
- Undergraduate/graduate startup grants
- Maker lab stipends or travel funds
- Pitch days or demo days run by innovation centers
- NIH/National Academy student research supplements

Case Snapshot: A BMTT-backed student startup bootstrapped $85K in non-dilutive awards across four pitch competitions and two institutional micro-grants *before* raising a cent of VC money.

5. THE ART OF EARLY-STAGE SCRAPPINESS

You don't need $5M to test whether your idea has legs. You need a kitchen table, a focused hypothesis, and maybe $5,000 in grant money or prize capital to prove it.

Capital-efficient early strategies:

- Use no-code tools instead of full-stack builds
- Run a physical prototype with off-the-shelf parts
- Join programs that offer resources in-kind (lab space, coaching, exposure)

- Partner with researchers or clinicians who already have access to infrastructure

Stretch every dollar by building toward validation, not perfection.

6. BUILDING A PIPELINE OF NON-DILUTIVE OPPORTUNITIES

Don't treat non-dilutive funding as one-off events. Treat it like a pipeline.

How to Systematize It:

- Create a shared tracking sheet of all relevant grants, fellowships, and pitch opportunities

- Assign deadlines and reminders 3–6 months in advance

- Prepare modular templates (biosketches, budget justifications, research aims) to reuse

- Reapply to the same funders across cycles (SBIR Phase I ⇨ Phase II, etc.)

BMTT Tip: Government reviewers often give detailed feedback. Losing round one? Treat it like a grant-funded roadmap.

7. WHEN TO RAISE AND WHEN TO KEEP BOOTSTRAPPING

Eventually, equity funding might make sense. But non-dilutive capital lets you **delay that decision** until your valuation is higher, your pitch is stronger, and your leverage is real.

Raise if:

- You need capital to meet an inflection point that grants can't fund

- There's urgency (e.g., clinical trials, first hires, exclusive licensing window)

- You've built enough momentum that VC conversations are already inbound

Keep bootstrapping if:

- You're still proving technical feasibility or market desirability

- You have access to grants and student resources

- Your idea benefits from careful iteration, not speed

CASE REFLECTION: THREE PATHS, ZERO DILUTION

1. The Grant Gambler (and Strategic Repeater)

A biomedical startup spun out of a graduate lab aimed to commercialize a neural monitoring device for stroke recovery. Rather than rushing into equity, the founder applied for an NIH SBIR Phase I with support from their PI and tech transfer office. They won ~$300K and, using reviewer feedback strategically, secured a Phase II award for ~$1.6M. When they later approached VCs, they did so from a position of strength and with meaningful validation.

Lesson: Treat grant feedback as a roadmap, not roadblock.

2. The Fellowship-Backed Launch

A solo founder developing a point-of-care diagnostic for pediatric respiratory illness won a fellowship designed for clinician-innovators (e.g., the NIH NIDDK Clinical Fellows). It covered a stipend, lab validation costs, and expert mentorship while allowing full IP control. Over two years, she validated prototypes with clinics and secured pilot agreements without surrendering equity.

Lesson: The right fellowship can catalyze early validation and maintain ownership.

3. The Scrappy Strategist

A student team with a wearable device targeting tremor suppression leveraged competitions and university micro-grants to raise over $80K. With a modular pitch deck, they reused biosketches and budget plans across applications. One advisor even granted lab access at no cost for prototype testing. This grassroots build enabled them to crowdfund a pilot and attract an industrial partner downstream.

Lesson: Non-dilutive wins stack. Smart repurposing of materials fuels compound momentum.

CLOSING THOUGHT: DON'T WAIT FOR A TERM SHEET TO BUILD

Some of the most compelling healthcare companies started with $2,000 and an overworked grad student.

Your idea doesn't need permission. It needs progress.

Non-dilutive capital lets you build that progress *on your terms.*

CHALLENGE: Build Your Non-Dilutive Capital Engine

You don't need VC to start. But you do need a system.

This challenge will help you identify immediate non-dilutive opportunities, organize them into a sustainable pipeline, and stretch every resource for maximum progress *before equity even enters the room.*

Step 1: List 5 Non-Dilutive Opportunities You Qualify For

These could include:

- SBIR/STTR grants
- Student startup competitions
- Innovation fellowships
- NIH supplements
- Pitch events at your university or region

If you're not sure, search for:

- "SBIR [your field] site:gov"
- "[University Name] innovation grants"
- "Undergraduate startup competition [year]"

Opportunity Name	Type (Grant / Fellowship / Competition)	Deadline	Potential Amount

Step 2: Draft One Modular Resource You'll Reuse Often

Pick **one** of the following and draft a first-pass version:

- 1-paragraph project summary

- 1-page biosketch or team overview

- Budget justification or milestone plan

- Generic pitch deck template (lightweight)

This becomes your plug-and-play asset. Update it once; use it everywhere.

Step 3: Identify a "Scrappy Sprint" You Could Fund With $5K

If someone handed you $5K tomorrow, how would you spend it to prove or de-risk something meaningful?

Examples:

- 3D print an early form factor

- Run 10 user interviews with incentives

- File a provisional patent

- Buy parts for a basic pressure or flow test

- Simulate a patient experience with a Figma prototype

Write it in one sentence:

"With $5K, we would _____ in order to test whether _____."

Step 4: Create a Funding Tracker Template

(Optional but powerful): Start a Google Sheet with columns like:

- Opportunity Name

- Source (agency/org)

- Deadline

- Required Documents

- Status (Not Started / In Progress / Submitted)

- Notes / Next Steps

Make it shareable if you have co-founders or collaborators.

Step 5 (Reflect): What Could You Build Without Giving Up a Single Share?

List **one milestone** you believe you could reach in the next 3–6 months *without raising equity.*

Then answer:

- What would you need (time, funds, access)?

- Which non-dilutive path(s) might help get you there?

Anchor This

Venture funding is one way to build. But it's not the only way. And it's rarely the best *first* way.

This challenge isn't about playing small. It's about building leverage on your terms, with your tools, and without asking for permission.

The goal isn't to avoid VC forever. The goal is to meet it on your terms when the time is right.

In the next chapter, you'll learn what VCs are *really* looking for so if and when you walk into that room, you'll be the one holding the leverage.

CHAPTER 32: WHAT DO VC'S REALLY WANT?

This chapter is not about how to charm VCs with flashy decks or clever metrics. It is about understanding what venture capitalists are actually *looking for* and *why they behave the way they do*.

Most founders walk into pitch meetings thinking investors are judging their product. They are not. They are judging your business. And more specifically, they are asking themselves one question: **Can this return the fund?**

Once you understand how VC funds operate, from how they're structured to how their timelines work, you'll stop trying to win every pitch and start aiming for the right pitch, to the right person, at the right time.

1. HOW VC FUNDS ACTUALLY WORK

The Basic Structure

- **Limited Partners (LPs):** These are the real money behind a VC fund. LPs are pension funds, endowments, family offices, or wealthy individuals who commit capital to a VC firm.

- **General Partners (GPs):** These are the investors you meet. They manage the fund, make investment decisions, and are responsible for generating returns.

- **Fund Lifespan:** Most VC funds operate on a 10-year cycle: 3-5 years to deploy capital, 5-7 years to see liquidity events.

Compensation Model

- **Management Fees (typically 2 percent per year):** Pays the GP team to operate the fund, regardless of performance.

- **Carried Interest (usually 20 percent):** The GPs take a cut of the profits *after* returning all capital to LPs. No profit, no carry.

Why It Matters to You

- GPs are incentivized to find companies that will exit with large multiples, **ideally 10x or more**.

- This means most VCs are not looking for solid businesses. They are looking for **outliers**.

- You might have a highly profitable company that does not excite them at all.

The GP across from you in a pitch meeting is balancing risk, portfolio theory, time pressure, and internal fund politics. They are not just asking, "Is this good?" They are asking, "Does this fit *my* model?"

2. CVCS VERSUS VCS

Not all capital is the same. Traditional venture capital and corporate venture capital look similar from the outside, but their goals and behaviors often diverge.

Traditional VC (financial funds)

- **Primary goal:** Return the fund through outsized financial outcomes.

- **How they help:** Company building, hiring, future fundraising, intros to other VCs, some customer access through networks.

- **What they optimize:** Ownership, follow-on reserves, path to a large exit within the fund's life.

Corporate VC (CVC)

- **Two common models:**
    - **Strategic CVC:** Optimizes for corporate value. Examples: product adjacency, new revenue lines, data access, supply chain security, strategic options.
    - **Financial CVC:** Optimizes for financial returns first, with strategic learning as a bonus. Often behaves like a traditional VC but with a corporate LP.
- **How they help:** Pilots with business units, distribution, credibility with industry buyers, regulatory and reimbursement insight, supply or manufacturing support.

Why this matters to you

- A strategic CVC may care more about pilot progress and fit with a business unit than near-term revenue.

- A financial CVC will often track traditional VC logic while still offering corporate access.

Common CVC term considerations

- **Right of first offer or refusal:** Can chill future acquisition interest if written too broadly. Narrow it by product line, geography, or timeframe.

- **Commercial rights or exclusivity:** Keep pilots non-exclusive unless compensated. Limit any exclusivity to a short pilot window, a narrow field of use, or both.

- **IP language:** Protect background IP and define foreground IP clearly. Avoid language that assigns broad rights to the corporate without explicit compensation.

- **Board role:** Prefer observer rights over a voting seat early, unless governance value is clear.

Syndication strategy

- A very strong pattern is VC lead, CVC follow, especially in health and medtech. The VC sets price and governance, the CVC adds distribution and pilot access.

- Some CVCs, however, do lead rounds, and when they do, they often bring their entire network with them. A leading CVC with conviction can pull in other corporates, traditional VCs, and even strategics in adjacent spaces.

- Example: Johnson & Johnson's venture arm is known for extremely thorough diligence and strategic syndication. When they lead, they rarely do it alone, and their reputation can de-risk the round for everyone else.

- The key is to understand whether the CVC in front of you typically plays lead or follow, and what that means for the rest of your fundraising strategy.

When to approach each

Choose CVC first when

- You need a pilot with a brand-name buyer to validate reimbursement or procurement.

- Distribution or integration with a specific platform is your fastest wedge.

- You are in a capital-intensive path where supply or manufacturing partnership is make-or-break.

Choose VC first when

- You need speed to hire and build, and you want a clean, independent cap table.

- You will raise multiple rounds quickly and need a lead with strong follow-on reserves.

- Your category has many acquirers, so you do not want to signal favored status too early.

Acquisition logic

- If you want optionality toward acquisition by the corporate, avoid broad rights of first refusal. Keep the door open for other buyers by narrowing rights by field, territory, or time.

Pilot success plan with a CVC parent

- Define the internal champion and the business unit now.

- Write a one-page pilot plan: success metric, timeline, owner, de-risking path after the pilot.

- Ask for a commercial next step in writing if the pilot hits the metric

---

※ **BMTT Tip:** Do not take strategic capital until a real champion inside the corporation is identified by name. If that champion changes roles, your deal can stall. Ask who owns the pilot internally and how success will be measured.

---

3. WHAT VCS ARE ACTUALLY LOOKING FOR

Forget the surface-level pitch questions. Here is what they are really evaluating.

1. Market Size

If your market cannot justify a billion-dollar outcome, they are likely not interested. Even if you can be profitable.

- Preferred: Huge TAM (Total Addressable Market), even if early.

- Acceptable: Mid-size TAM with strong adjacent markets or an entry into a bigger space.

2. Team

They are judging your founding team's ability to execute, adapt, and attract talent. Not your résumé. Your *resilience and clarity*.

- "Is this team a magnet for other top talent?"

- "Will they listen without ego?"

- "Can they sell, recruit, and build under pressure?"

3. Timing and Traction

You do not have to be post-revenue, but you do need to show that something is working: user growth, demand, partnerships, or strong validation from users.

- If you are pre-traction: they better believe the timing is right.

- If you are post-traction: show growth velocity or evidence of a future inflection point.

4. Exit Potential

If they cannot envision a clear exit path within 7 to 10 years, they will likely pass.

- Will a large player acquire you?
- Are there recent comps with successful exits?
- Could you go public?

4. READING BETWEEN THE LINES IN PITCH MEETINGS

What It Sounds Like vs. What It Means

What They Say	What They Likely Mean
"Interesting. Let us keep in touch."	They are not convinced. You are not a priority.
"How much are you raising?"	They are testing if your strategy aligns with your stage.
"What does success look like for you?"	They are evaluating if you understand VC-scale thinking.
"Who else is in the round?"	They are assessing social proof and lead-follow dynamics.

Green Light Moments

- They ask about specific uses of funds and suggest allocations.

- They volunteer to make introductions without you asking.

- They follow up within 24–48 hours with clear next steps.

Case Study: Real VC Engagement
A medtech founder pitched a platform to improve post-op pain management. The VC asked for a follow-up meeting and introduced them to three surgical groups for pilot conversations. That is not curiosity. That is a VC beginning to lean in, testing demand and doing diligence on your behalf.

CULTURE AND GEOGRAPHY: HOW APPROACH SHIFTS

Investors are people first. Culture, communication style, and geography change how you read a room and how you follow up. Treat these as tendencies, not rules.

United States

- **Style:** Optimistic, fast, relationship-first. Will often say yes to another meeting, even if conviction is low.

- **Phrases:** "Interesting" or "Let us keep in touch" can signal a soft no unless paired with concrete next steps within 24 to 48 hours.

- **Approach:** Ask for a clear action: pilot intro, partner call, diligence doc request.

United Kingdom

- **Style:** Polite, measured, preference for preparation and precision.

- **Phrases:** "Not for us at the moment" usually means no. "Do share updates" can be genuine if they specify what to send.

- **Approach:** Lead with data and references. Avoid hype. Follow up with a concise monthly update.

Germany

- **Style:** Direct, detail-driven, value clarity and engineering rigor.

- **Phrases:** "Interesting" often means genuinely worth deeper review. Expect specific questions and requests for technical validation.

- **Approach:** Bring test data, standards compliance plans, manufacturing depth, and a clear regulatory map.

Israel

- **Style:** Very direct, fast, founder-operator mindset. Comfortable with risk and iteration.

- **Phrases:** "Cut to the chase, what do you need" is common. If they like it, you will know quickly.

- **Approach:** Lead with problem severity, traction signals, and the unfair advantage. Be ready to whiteboard the plan.

France and the Nordics

- **Style:** Thoughtful, design and systems oriented, consensus friendly.

- **Approach:** Map the ecosystem, partnerships, and sustainability logic. Emphasize product quality and long-term moats.

Reading between the lines everywhere

- If interest is real: they volunteer intros, ask allocation questions, or request a specific diligence artifact.
  If interest is light: they ask you to "keep them updated" without a concrete request or date.

※ **BMTT Tip:** Mirror the investor's style without losing your own voice. If they are precise and data-first, tighten your answers. If they are rapid and brainstorming, get to the point, then co-create next steps.

5. WHEN YOU SHOULD (AND SHOULDN'T) RAISE VENTURE CAPITAL

Raise Venture Capital If:

- You are solving a problem in a market that needs *fast* scaling to beat competition.

- Your product requires deep upfront capital (e.g., clinical trials, manufacturing scale).

- You are ready for institutional pressure (reporting, board seats, and timelines).

Avoid Venture Capital If:

- You want long-term control and slower, sustainable growth.

- Your product serves a niche market that does not justify a massive return.

- You have access to alternative funding (e.g., grants, revenue, strategic partners).

✳ **BMTT Insight:** Some of the most successful health startups are VC-backed. Some of the most *stable* are not. Know what you want your life to look like five years from now and then decide whether VC belongs in that picture.

6. DO YOUR VC DUE DILIGENCE BEFORE OUTREACH

Strong outreach begins with strong targeting. Do not ask investors to educate you on their focus. Show you did the homework.

Fund fundamentals to verify

- **Stage:** Pre-seed, seed, Series A, later. Check actual first-check behavior, not just website language.

- **Check size and ownership targets:** Typical first check, target percent, reserve strategy for follow-ons.

- **Sectors and theses:** Current themes, partner specialties, recent deals that match your wedge.

- **Geography and founder profile:** Regions they actively support, any focus on student or technical founders.

- **Fund vintage and dry powder:** New fund often equals higher deployment velocity. Late-vintage fund can be slower or focused on follow-ons.

- **Conflict risk:** Overlap with portfolio companies. Some will still engage if the overlap is adjacent and disclosed.

Signals the time is right

- Partner wrote publicly about your space in the last 6 to 12 months.

- Portfolio company is a potential channel or integration for you.

- They just led a round at your stage in an adjacent problem area.

How to research quickly

- Firm and partner pages, partner posts, podcasts, and talks.

- Your Crunchbase Pro account for portfolio, check sizes, round timing.

- Founder backchannels: ask two portfolio CEOs what working with this fund is actually like.

Build a one-page fund brief before every outreach

- Partner name, thesis quotes, last three relevant investments.

- Check size and ownership targets.

- Why your wedge fits their pattern.

- Your ask for the first call.

※ **BMTT Tip:** You are not pitching a firm. You are pitching a partner who will advocate inside Monday partner meeting. Target the human who has shown interest in your exact wedge.

7. TANGIBLE OUTREACH SKILLS

Your five-sentence cold email

Subject line: Company name and one-line wedge. Example: "NeuroWave: wearable for early stroke detection, validated in pilot study."

Who you are in one line: student founder, lab, or spinout, plus location.

Why now: traction signal or timing insight. Example: pilot LOI, regulatory gate cleared, cost curve change.

Why them: connect to their thesis or portfolio. One sentence with specifics.

The ask: a 20-minute calibration call next week, and a link to a one-pager.

What to attach

A crisp one-pager or five-slide teaser. Save the long deck for the call.

Links to a 60 to 90 second demo or data snapshot if relevant.

No NDA. Offer a light data room only after mutual interest.

Warm introductions the right way

Founders are often the best bridge to a VC. A partner may ignore a cold email but will almost always read a note forwarded by a portfolio CEO they trust.

Identify two or three portfolio companies in your space or at your stage, then reach out to those founders. Do not ask for an intro immediately; ask for advice on how they worked with the VC, what diligence was like, and what they wish they had known. If the conversation goes well, they may offer to connect you.

When a founder agrees to intro you, send them a tight blurb they can forward. Include the same five sentences and the ask. Make it effortless to introduce you.

Remember: founders protect their investor relationships. Respect their time, and show that you did homework on both their company and the VC before asking.

Follow-up cadence

If no reply in one week: send a short bump with one new fact.

If still quiet: move to monthly updates with two bullets and a single request.

First call goals

Confirm fit on stage, check size, and timing.

Identify the partner's champion interest area and what they need to see next.

Leave with one concrete next step and a date.

BMTT Tip: Treat the first call as a discovery interview. Ask what milestones would put you in their strike zone in the next 60 to 90 days, then reflect those back in your follow-up note.

CASE REFLECTION: SEEING IT FROM THE VC SIDE

1. The "No" That Meant "Not Yet"

A digital diagnostics startup pitched a seed-stage health tech VC and heard, "We love the space... keep us posted." At first, they took it as rejection. But they later realized they lacked both TAM clarity and a clear path to reimbursement. After refining their go-to-market strategy, adding mock CPT pathways, and reframing their TAM with adjacent use cases, they re-pitched six months later and got a check.

Lesson: VCs are rarely judging your current value. They're judging your current story. Make the story better, and they might change their mind.

2. The Niche They Couldn't Justify

A student team building a powerful neuro-rehab device had strong pilot data and deep patient validation, but the core market was too niche to excite institutional VCs. One investor explained it bluntly: "*Even if you dominate this space, it's not a fund-returner.*" Rather than force a raise, they partnered with a strategic, non-VC-aligned fund and focused on milestone-based grants. Three years later, they sold the IP to a rehab tech company.

Lesson: VC money is only right if you're solving VC-shaped problems. You can still build a winning company without chasing unicorn math.

3. The VC Who Jumped In Early Because the Exit Was Obvious

A founder pitching a digital therapeutic didn't have revenue but had lined up LOIs from three major hospital networks. The VCs in the room saw more than traction; they saw acquisition logic. A clear need, obvious buyer, and validated interest got them early backing despite a barebones prototype.

Lesson: Sometimes it's not about scale. It's about story. If they can clearly picture a liquidity event, they'll often take the risk.

CLOSING THOUGHT: PITCH TO LEARN, NOT JUST TO RAISE

Your first few VC conversations are rarely about closing money. They are about learning how your company is perceived from the other side.

You are not just fundraising. You are pressure-testing your business model in the eyes of someone who sees thousands of pitches and only bets on a few.

Respect their time. Understand their incentives. And remember: it is not rejection if they are simply playing a different game.

CHALLENGE: Pitch From Their Side of the Table

This chapter didn't just tell you how VCs think; it handed you their playbook. Now, it's your turn to flip the perspective.

Your challenge is to step *into the shoes of an investor* and evaluate your own venture the way they would.

Step 1: Answer the Real VC Questions

Answer each question as honestly and clearly as possible (no pitch language, no fluff).

1. **Can this return the fund?**
   ⇨ Based on our market size and exit potential, could this venture realistically lead to a $100M–$500M outcome?

2. **Is this a billion-dollar market (or a bridge to one)?**
   ⇨ What is the real TAM (Total Addressable Market), and is it growing?

3. **Do we have VC-scale traction or velocity?**
   ⇨ What traction do we have today (users, partners, data)? If not much, why is the timing still right?

4. **Would we invest in us?**
   ⇨ Does our founding team look VC-backable: complementary, ambitious, resilient, and ready to hire and scale?

Step 2: Create Your One-Slide VC Snapshot

Using your answers above, mock up a single slide (just in your notes or on paper) that a VC might build about you after a meeting.

It should include:

- Startup name + one-liner
- Founding team quick bio
- Market opportunity (TAM + trends)
- Traction (current + projected milestones)
- Exit logic (potential acquirers or IPO case)

This helps you see what narrative they're walking away with. Does it align with the story you want to tell?

Step 3: Build a Fund Fit Scorecard

Use this scorecard to test whether a fund is worth your time. Rate each category 0–5 and total out of 30.

- Stage fit
- Check size fit
- Sector thesis match
- Partner champion identified
- No direct conflict with portfolio

- Timing and fund vintage fit

Score 24 or above: green light. Score 18 to 23: nurture. Below 18: monitor and update, but do not spend cycles now.

Step 4: Draft and Test Outreach

Write your five-sentence cold email using the template in this chapter. Then, write a second version as if you were asking for a warm intro through a portfolio founder. Share both with a peer or mentor and ask which feels more compelling. Iterate until both versions feel sharp, specific, and respectful of the reader's time.

Step 5: Identify Founder Intro Paths

Pick two funds that interest you and identify one portfolio founder at each you could reasonably approach. Reach out for advice, not an intro. Ask how they worked with the VC, what diligence was like, and what they wish they had known. If the conversation builds trust, they may offer the intro themselves. Document what you learned regardless of outcome.

Step 6: Define One VC-Aligned Move for the Next 90 Days

Even if you're not raising VC now, pick *one thing* you could do that would make your startup more fundable to them later.

Examples:

- Interview 10 more users to better understand pain points and buying logic

- Refine your TAM calculation based on adjacent market dynamics

- Add a technical or commercial co-founder to complete the team

- Land an LOI or pilot to show pre-revenue demand

- Improve your investor-facing Lean Canvas

Write it down, and schedule 1 hour this week to start it.

Anchor This

You don't need to *become* a venture-scale company. But you do need to understand how they think, because whether you pitch this startup, your next one, or just want to build something worth betting on, it helps to know what game they're playing.

This challenge isn't about chasing VC. It's about learning how to *own the room* when it matters and choosing, with clarity, whether you ever want to walk into it again.

Part XII

Building Forward

You've made it this far. By now, you've seen the shape of the journey: from defining problems and mapping landscapes, to testing solutions and crafting a pitch, to understanding funding and translating your work into strategy. Each step has been framed with one purpose in mind: making sure that student innovation doesn't die quietly on a shelf but instead has a pathway to grow, adapt, and thrive.

This closing chapter is not another checklist, nor a "challenge" like the others. Instead, it is an invitation to zoom out, gather what you've built so far, and set your sights on what comes next.

1. From Surviving to Sustaining

Early-stage building is often about survival: Will this idea work? Will anyone care? Will I find the time, the money, or the team? You've wrestled with those questions already. The path forward is about moving from *just surviving* to *sustaining*.

That shift begins with clarity on three fronts:

- **Metrics that Matter**: You've learned to distinguish vanity metrics from meaningful ones. As you move forward, focus on traction signals that prove both value and usability, not just activity.

- **When to Double Down vs. Pivot**: Every builder eventually faces the moment where the original plan no longer holds. The key isn't avoiding pivots but knowing when they're strategic, not reactive.

- **Balancing Vision with Iteration**: The best founders hold onto the original spark but are ruthless about improving execution. You've practiced this balance in your problem statements, prototypes, and canvases. Keep practicing it as you scale.

2. Culture as Strategy

By now you've seen that startups are not built by tools or templates alone. They are built by people, and the culture those people create will either accelerate or suffocate progress.

As you look ahead:

- **Protect the Core**: Keep the mission visible. Students, mentors, and partners will come and go, but the "why" behind your effort should never blur.

- **Build with Transparency**: From term sheets to team roles, clarity beats confusion. Startups collapse under ambiguity more often than under competition.

- **Grow a Community, Not Just a Company**: The projects that endure become more than companies; they become networks. Customers who feel invested, mentors who stay engaged, peers who share your wins. Think beyond transactions, build relationships.

3. Learning as a Permanent State

This book has emphasized iteration at every turn: sharpening your problem statements, refining prototypes, listening harder in customer discovery. That doesn't end when you raise money or launch a product.

The founders who last are the ones who keep harvesting lessons.

- **Wins Teach as Much as Losses**: When something works, don't just celebrate. Ask why it worked and how to repeat it.

- **Codify Learning Into Your Cadence**: Don't wait for crises. Build in regular time to reflect, capture insights, and translate them into action.

- **Beware the Success Plateau**: Early wins can create complacency. Treat them instead as signals that your system is working and then push it further.

4. Building Forward

You began this book with the idea that you didn't need to be "ready" to start. You close it with a reminder: you're never really finished. Every company, every venture, every system you build is simply a step toward the next horizon.

Building forward means three things:

- **Carrying What You've Learned**: The tools here are portable. Whether you launch one startup or ten, whether you work in healthcare or another industry, the process of disciplined problem discovery, validation, and execution carries with you.

- **Expanding Who You Build With**: Start with classmates and mentors, then expand. Building forward is about widening the circle until your idea touches people and partners you didn't know you needed.

- **Anchoring to the Bigger Picture**: The end goal isn't just a successful company. It's creating a world where student innovation is no longer wasted, where fresh perspectives consistently turn into real solutions.

Closing Note

If you've read this far, you already believe in the possibility of building forward. You've invested the time to learn frameworks, the discipline to practice them, and the courage to imagine yourself not just as a student, but as a builder.

There is no final challenge here because the next step is yours to design. This is the end of the playbook, but it is not the end of your story. The ideas you carry are only waiting for the chance to be built.

So go build forward.

Challenge Index

Chapter 3 – Exploring your Builder Identity
Reflect on your motivations, strengths, and long-term goals as a builder; define what kind of innovator you want to be.

Chapter 5 – Your First 20 Problems
Observe real-world problems and list at least 20 that feel meaningful, actionable, and worth solving.

Chapter 6 – Sharpening the Signal, Not the Scope
Filter your ideas for clarity and specificity, ensuring you're pursuing a clearly defined problem rather than a vague space.

Chapter 7 – From Raw List to Ready Shortlist
Narrow your ideas from many to a focused few that are feasible, aligned with your skills, and scoped for near-term action.

Chapter 8 – The Landscape Gauntlet
Conduct a landscape review to assess competition, adjacent efforts, and whitespace before committing further.

Chapter 9 – Your Problem, In the Wild
Run customer discovery interviews or field research to understand your target users and their lived experience with the problem.

Chapter 10 – Building a Market That's Ready, Not Just Big
Validate that a large market exists and that people actively seek solutions, not just that the problem is theoretically big.

Chapter 11 – Generating with Discipline, Not Desire
Generate multiple possible solutions, but constrain yourself with real-world limitations and user feedback to guide your ideas.

Chapter 12 – Co-Create with Constraint
Bring users into your ideation process to shape early prototypes with grounded insights and practical constraints.

Chapter 13 – Build to Learn, Not to Impress
Build early prototypes purely to learn, not to impress; identify what assumptions you're testing and design accordingly.

Chapter 14 – Earn Your MVP
Develop a functional MVP that proves one critical insight: not a polished product, but a focused experiment.

Chapter 15 – Run Your First Real Campaign
Plan and execute your first structured test campaign... whether for user interest, clinician feedback, or stakeholder reaction.

Chapter 16 – Make the Call
Use your feedback and results to choose: iterate, reframe the problem, or move on from the idea entirely.

Chapter 17 – Secure Your Edge
Analyze whether you have a protectable edge (technical, data-driven, IP-based, or other) and how to communicate it.

Chapter 18 – Walk the Five-Pillar Gauntlet
Evaluate your startup using a five-pillar framework: value, feasibility, defensibility, scale, and alignment.

Chapter 19 – Build the Clinical Compass
Interview clinicians or experts to map your solution's clinical relevance, points of intervention, and usage pathway.

Chapter 20 – Build Your Business Model Blueprint
Sketch out your business model (who pays, who uses, and how you create and capture value).

Chapter 21 – Define Your Commercialization Path
Outline how your solution would move from concept to market: through academia, licensing, startups, or partnerships.

Chapter 22 – Build Your Startup's Legal Backbone
Understand IP ownership, contracts, and entity structures and then draft a simple legal foundation for your venture.

Chapter 23 – Turning Planning into Proof
Convert your assumptions and strategy into a preliminary operating plan: who will do what, when, and with what tools.

Chapter 24 – Turn Uncertainty into Actionable Data
Design small, controlled experiments to reduce uncertainty and gather early proof or disproof of key assumptions.

Chapter 25 – Build Your Lean Canvas
Use the Lean Canvas template to lay out your full venture hypothesis in one structured visual summary.

Chapter 26 – Craft Your Investor-Ready Pitch Deck
Build a professional, minimalist, and clear pitch deck aligned with investor logic and aesthetics.

Chapter 27 – Build the Team that Builds the Startup
Find mentors/advisors and evaluate your founding team's fit, culture, and gaps, and define how you'll grow and lead under pressure.

Chapter 28 – Draft Your Funding Fit Snapshot (Without Making a Decision Yet)
Research available funding mechanisms and assess which ones might be a fit for your startup stage and sector.

Chapter 29 – Know What You're Signing Before You Even Sign it
Learn the key deal terms that show up in early-stage term sheets and SAFEs and what they mean for you long-term.

Chapter 30 – Map Your Milestones, Not Just Your Money
Map your planned capital needs against your development milestones to determine when and how much to raise.

Chapter 31 – Build Your Non-Dilutive Capital Engine
Draft a plan to pursue non-dilutive funding opportunities including grants, competitions, and pilot programs.

Chapter 32 – Pitch From Their Side of the Table
Reframe your pitch from the investor's perspective: what they need to see, hear, and believe to fund you.

While I hope this playbook results in exited startups
and lives saved, my deeper hope is that it convinces
you of one thing:
you can do this, and the time to start is now.
If you ever have questions or doubts,
I invite you to reach out on LinkedIn.

Parker Brewster

Made in the USA
Coppell, TX
13 November 2025

63058704R00262